Fuzzy Logic Controllers and Applications

Edited by P. Balasubramaniam
and N. Ramesh Babu

Published in London, United Kingdom

Fuzzy Logic Controllers and Applications
http://dx.doi.org/10.5772/intechopen.1004663
Edited by P. Balasubramaniam and N. Ramesh Babu

Contributors
Chiheb Reddouane, Fatima Ouzayd, Hamid Ech-cheikh, Israa Salih H. Alajab, Kirenia de la Caridad Viamontes Rabelo, Luis Benigno Corrales Barrios, Manal Tamir, Mehdi Dalir, Mohammed Osman Hassan, Namal Rathnayake, Nooshin Bigdeli, Tuan Linh Dang, Upaka Rathnayake, Yainier Otero Rosell, Yukinobu Hoshino

First published in London, United Kingdom, 2025 by IntechOpen
IntechOpen is the global imprint of INTECHOPEN LIMITED, registered in England and Wales, registration number: 11086078, 167-169 Great Portland Street, London, W1W 5PF, United Kingdom

For EU product safety concerns: IN TECH d.o.o., Prolaz Marije Krucifikse Kozulić 3, 51000 Rijeka, Croatia, info@intechopen.com or visit our website at intechopen.com.

British Library Cataloguing-in-Publication Data
A catalogue record for this book is available from the British Library

Fuzzy Logic Controllers and Applications
Edited by P. Balasubramaniam and N. Ramesh Babu
p. cm.
Print ISBN 978-0-85014-874-9
Online ISBN 978-0-85014-873-2
eBook (PDF) ISBN 978-0-85014-875-6

Meet the editors

P. Balasubramaniam has worked as a Senior Professor in the Department of Mathematics at The Gandhigram Rural University, Gandhigram, India, since 1997. He has published over 353 research papers in various SCI journals with impact factors with Scopus H-index 50 and Google Scholar H-Index 56. His research topic includes fuzzy logic, control systems, and machine learning. He serves as a reviewer in many SCI journals and a member of the editorial board of a few Journals. He is the Editor-in-Chief of Modern Instrumentation Journal and Associate Editor of Neural Processing Letters and Network: Computation in Neural Systems. He has been recognized as one of the Top 2% of scientists in the field of AI & Image Processing since 2019.

N. Ramesh Babu is currently working as an Assistant Professor (Grade 1) in the Department of Mathematics, School of Advanced Sciences, Vellore Institute of Technology, Chennai, India. He received his Ph.D. in Mathematics from The Gandhigram Rural Institute (Deemed to be University) in 2023, focusing on synchronizing chaotic dynamical systems and their applications. His research interests include dynamical systems, image processing, cryptography, and neural networks. Dr. Ramesh Babu has received multiple accolades, including the Sydney R. Parker and M. N. S. Swamy Best Paper Awards in 2022. As a member of a professional society like the Ramanujan Mathematical Society, he continues to contribute to advancements in mathematical modelling and applied sciences.

Contents

Preface

In the ever-evolving field of control systems, pursuing more adaptive, intelligent, and robust solutions has led to the development of innovative techniques that transcend the limitations of traditional control methods. One such ground-breaking approach is fuzzy logic, which has gained significant prominence for its ability to handle uncertainty and imprecision—concepts inherently present in real-world systems. The application of fuzzy logic to control systems, often encapsulated within Fuzzy Logic Controllers (FLCs), has opened new avenues for solving complex problems in various industries.

This book contains two sections, "Fuzzy Logic Controllers" and "Applications of Fuzzy Logic". It aims to provide two fuzzy logic control-related chapters and three application chapters. Combining theoretical foundations with real-world applications presents both novice and experienced engineers with a deep understanding of how fuzzy logic can be utilized to achieve improved performance in systems that are difficult to model with conventional methods.

Applying fuzzy logic to control systems is an exciting and rapidly advancing field, and this book seeks to illuminate its potential. We hope it inspires further exploration and innovation, encouraging the next generation of engineers to leverage fuzzy logic to solve the complex challenges of tomorrow's technology-driven world.

P. Balasubramaniam
Senior Professor of Mathematics,
The Gandhigram Rural Institute (Deemed to be University),
Tamil Nadu, India

N. Ramesh Babu
Assistant Professor of Mathematics,
School of Advanced Sciences,
Vellore Institute of Technology,
Tamil Nadu, India

Section 1

Fuzzy Logic Controllers

Chapter 1

A Fuzzy Logic Controller for Enhancement of the Power Grid Single Contingency Management

Israa Salih H. Alajab and Mohammed Osman Hassan

Abstract

Over the past two decades, prodigious prosperity has occurred in the utility of fuzzy logic controllers in power systems and power electronic applications. One of the major issues for the power system engineer is to monitor and rank the network contingencies related to an outage of one or more elements on the power network. Contingencies denote a single contingency (N-1) for the outage of one component or multiple contingencies for more than one element. When contingency occurs, the system may operate out of its stability margin and become insecure. Hence, it is necessary to manage that according to their impact on the network. The chapter discusses one of the cutting-edge applications of fuzzy logic control systems on electrical power systems by presenting an intelligent approach of fuzzy logic to rank the N-1 in the Power Grid. The intrinsic objective of this study is to develop a MATALB fuzzy logic model for ranking contingencies on the abridge network of Sudan's national grid and compare this ranking with that which was chiefly accomplished using NEPLAN software. The results on applied partial network confirmed that the Fuzzy Logic applied algorithm is an effective and proper ranking process for the grid single contingencies (N-1).

Keywords: fuzzification, defuzzification, membership function, severity indices, national Sudanese grid (NSG), violation, contingency ranking

1. Introduction

Fuzzy logic controllers (FLCs) have produced a great deal of interest in several utilizations and have been imported into power electronics. They possess multiple advantages compared with the conventional PI controllers, which can operate with imprecise input and handle nonlinearity, in addition to their higher durableness than PI controllers, also they do not need an accurate mathematical model [1].

Fuzzy logic control has expressed a momentous sector of fuzzy reasoning, which is illustrated by control theory based on mathematical models of the open-loop progress to be controlled [2].

The influence of the element outage on the power system, when the rest of the system is in service, is called a contingency study [3]. The security operation of any power grid is a highly significance issue; hence, the contingency ranking is an integral section in the power networks security assessment [4]. The main target of contingency capturing and ranking is to hastily and meticulously shortlist critical contingencies from a massive list of reliable contingencies and rank them depending on their severity for advanced conclusive analysis [5].

In recent years many researchers have going ahead on the contingency analysis, Alam proposed a method to determine the power system contingency ranking by using voltage deviation indices and power deviation indices [6], and Christie and Talukdar presented a consolidation technique for evaluation contingencies depends on pundit systems for contingency election and interpretation, but the execution depends on the knowledge base which becomes so slow due to the knowledge base castles [7]. The using of intelligent systems in the contingency ranking had a wide range For instance, Mohsena et al., provide an integrated algorithm for contingency ranking by taking into account the indices of voltage violation, line flow violation, locational marginal price and congestion cost using the neural network to determine the main parameters of power system [8]. Fuzzy logic was also used in contingency studies by Ramesh et al. to remove the mask effect after ranking contingency by using apparent power overloading and voltage violations. In this chapter, a fuzzy logic system has been used to accomplish the process of ranking for single contingencies [9].

As shown previously, although there are different studies discuss the fuzzy logic in the contingency analysis, no one of them applies the fuzzy logic system in ranking the single contingencies in the Sudan national grid which is the novelty of this study which contributes to advances in fuzzy logic system by applying this technique in a real national grid to rank the single contingencies.

Sections of this chapter may contain the following: the methodology of research in section 2 and overall reviews on the implemented network characterization in Sections 3. Section 4 represented the simulation of a fuzzy logic control system with its two processes (fuzzification and defuzzification). Section 5 illustrated the fuzzy logic simulation results analysis and discussion compared to those contingency raking obtained from NEPLAN software contingency in which both subsections discussed the results of single contingency ranking from NEPLAN software and results analysis of fuzzy logic system contingency ranking. The later sections is the conclusion and recommendations which they summarized the chapter results. Finally, the references for the previously published literature have been included.

2. Methodology

According to the objectives of the study, the base requirement is to accomplish the contingency ranking by using the fuzzy logic control system, and the process relies on two base electrical amounts: the first is voltage and the second is megawatt loading which is used as inputs for fuzzy logic system that both of voltage profile and line power flow had a created membership function of trapezoidal type "trapmf" on the Mamdani model membership at fuzzy interface system (FIS) toolbox on *MATLAB*. The output of the (FIS) is the severity indices of voltage profile (SIVP), and the severity indices of line flow (SILF). The algebraic summation of (SIVP) and (SILF) typify a composite index (CI) which the ranking process depends on. The most severe contingencies have a higher (CI) and vice versa. This process is illustrated in **Figure 1** which represents the generation of CI.

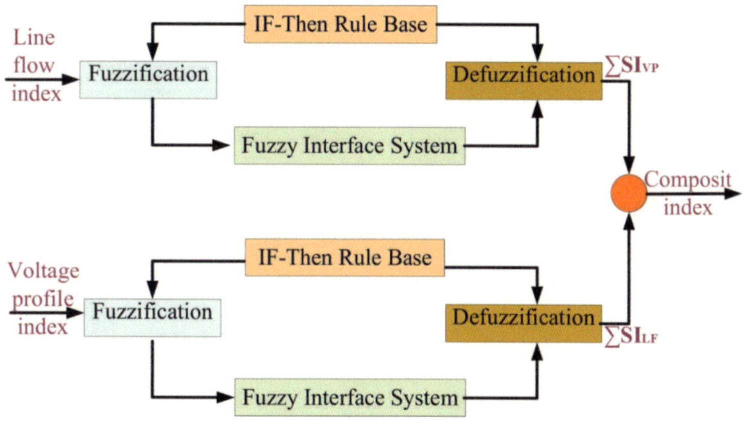

Figure 1.
Composite index identifying based on fuzzy logic system [10].

For reliability purposes, the results of contingency ranking were compared with those that were investigated at normal load in NEPLAN software. NEPLAN (Network Planning) is a network simulation program that acts multi-tasks of the power system analysis such as (load flow, short circuit, capacitance location, contingency analysis, etc.). Ranking of contingency at NEPLAN depends upon the concept of violation level that an element outage causes voltage violation on remaining network elements, if the violation is above (100%), the system is insecure and the contingency according to the violated element has been ranked as most severe.

3. The implemented network characterization

A partial grid of the national Sudanese grid is implemented in this study, as shown in **Figure 2**. The simplified eighty-two (82) bus-bars operate at four multifarious voltage levels (500, 220, 110, and 66 kV), which form the main part of the transmission system in the Sudan electrical grid. The total number of transmission lines on the network had been summarized as eighty-one (81) which connected the main supplied areas.

The data was taken at normal load conditions where the network simplified to total generation and total loading as: (1603.517 MW, and 227.008 MVAr) and (1553.3 MW and 1083.3 MVAr) respectively.

The total generation in the grid is generated from different sources (Hydro generation, Solar, and most are thermal power plants in addition to the interconnection tie line feeders from both Ethiopia and Egypt).

4. Simulation of fuzzy logic control system

As clarified previously on the methodology, the membership functions had already been created for voltage profile and load flow indices as can be shown in **Figures 3** and **4**.

The outputs of the two membership functions represent the severity indices of (SIVP) and (SILF) which can be shown in **Figures 5** and **6**.

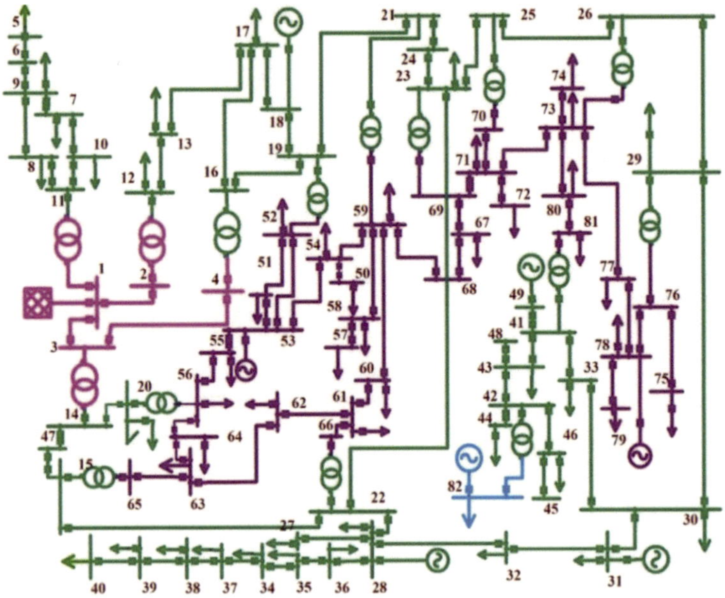

Figure 2.
Single-line diagram of the implemented partial Sudan grid [11]. ■ *500 kV,* ■ *220 kV,* ■ *110 kV,* ■ *66 kV.*

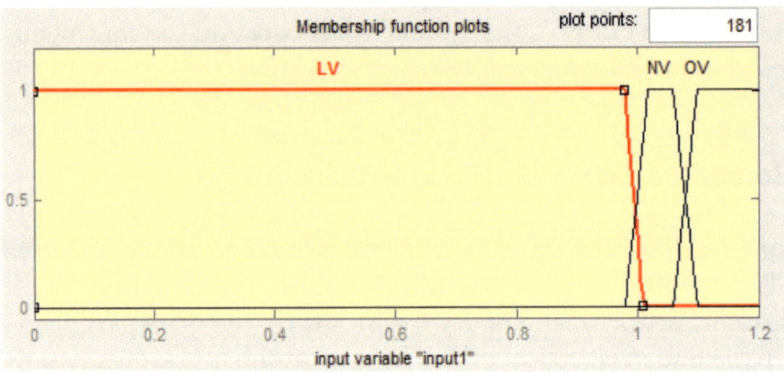

Figure 3.
The created membership function of the voltage profile [11].

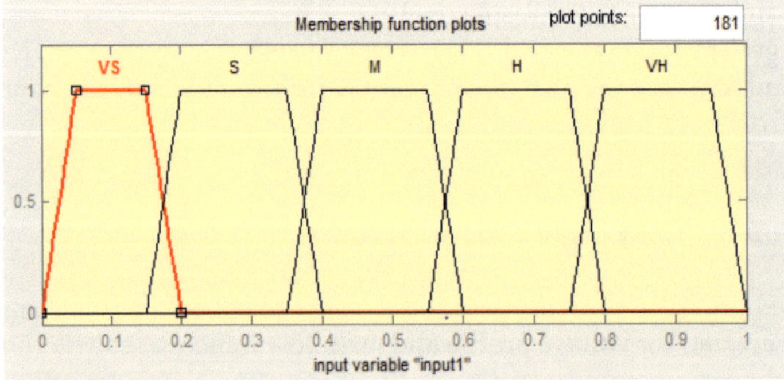

Figure 4.
The created membership function of line flow.

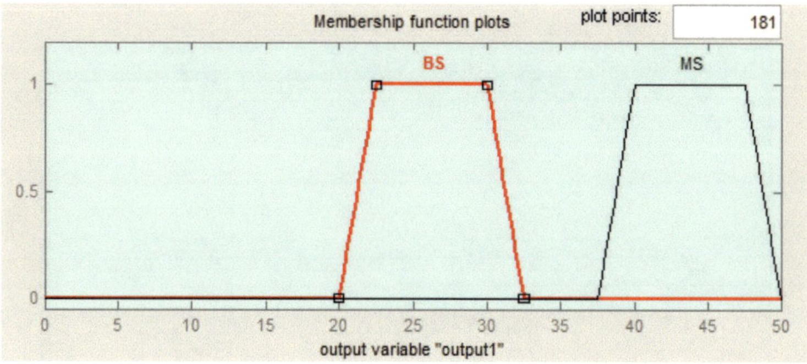

Figure 5.
The severity index of voltage profile membership function.

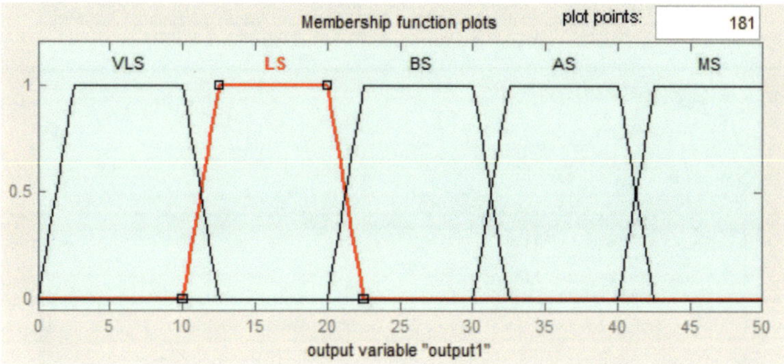

Figure 6.
The severity index of the line flow membership function.

4.1 Voltage profile and line power flow fuzzification

The fuzzification which already had been processed through the Mamdani Functions model (MFs), the IF-then rules were set as in **Table 1**.

Table 1 can be implemented on the rule editor at (FIS) on MATLAB as shown in **Figures 7** and **8**.

The If-then rules formulated in **Figures 7** and **8** can be depicted clearly on the rule viewer at FIS as shown in **Figures 9** and **10**.

Input	**Voltage profile**	**LV**		**NV**		**OV**
	Line flow index	**VS**	**S**	**M**	**H**	**VH**
Output	**SI_{VP}**	**MS**		**BS**		**MS**
	SI_{LF}	**VLS**	**LS**	**BS**	**AS**	**MS**

Where: LV= Low Voltage, NV= Normal Voltage, OV= Over Voltage, MS= More Severe, BS= Below Severe, VS= Very Small, S = Small, M= Medium, H= High, AS= Above Severe, VH= Very High, LS= Less Severe, VLS= Very Less Severe.

Table 1.
Fuzzification rules of voltage profile and line flow.

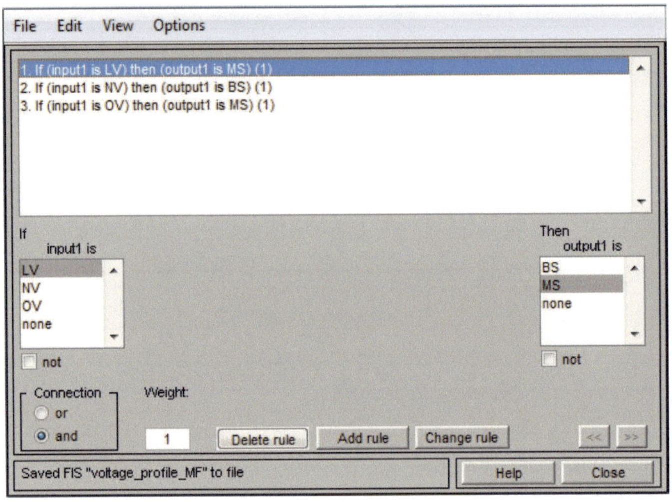

Figure 7.
The rules editor of voltage profile MF [11].

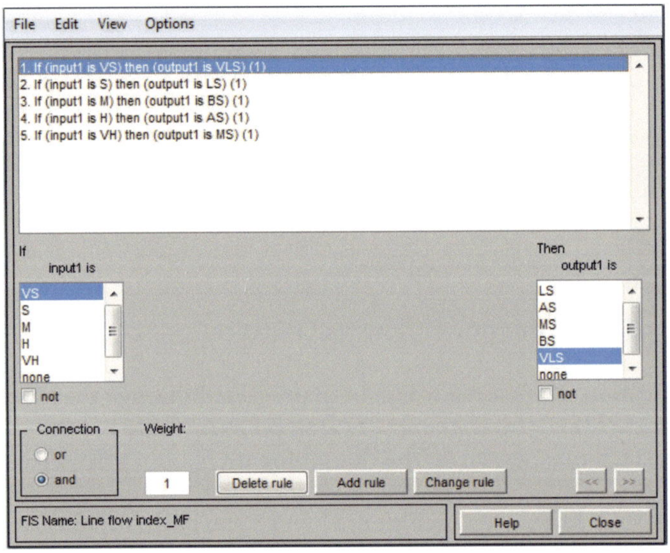

Figure 8.
The rules editor of line flow MF [12].

4.2 Severity indices defuzzification

Defuzzification of the output is the process in which the severity indices of voltage profile and line flow (SIVP and SILF) are illustrated as numbers two different membership functions with *linguistic variables* of voltage profile data and line flow data of the network [15], on the other hand, *linguistic values* of these two membership functions represent both of severity index of voltage (SIVP) and severity index of line flow (SILF). This defuzzification process can be interpreted clearly in **Figures 11–14**.

As mentioned earlier in the methodology section, the ranking of the fuzzy approach is based on the composite index (CI) [17]. The CI expresses the total

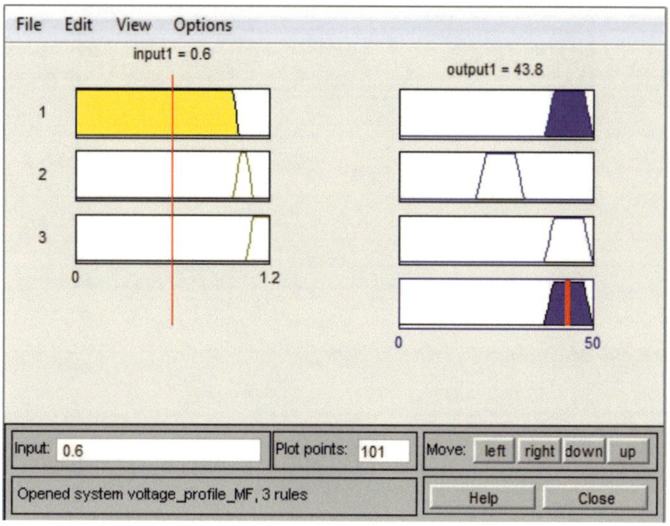

Figure 9.
The rules viewer of the voltage profile membership function [13].

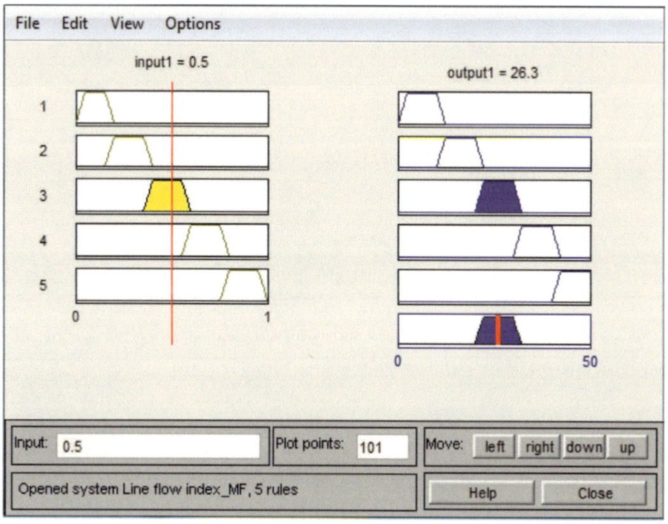

Figure 10.
The rules viewer of the line flow membership function [14].

Figure 11.
Voltage profile fuzzy logic controller block [10].

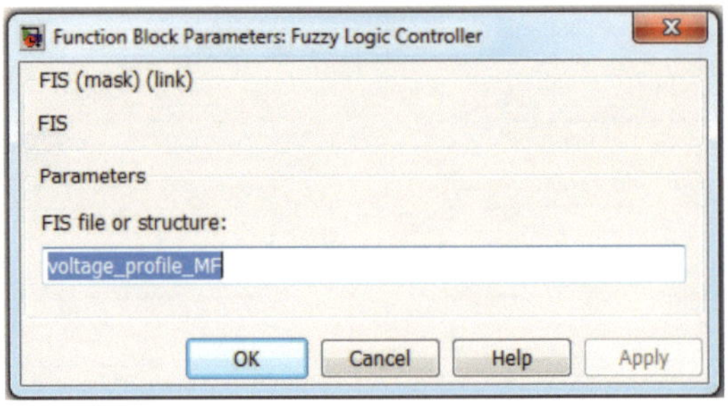

Figure 12.
Inserting of voltage profile MF on fuzzy logic control block [16].

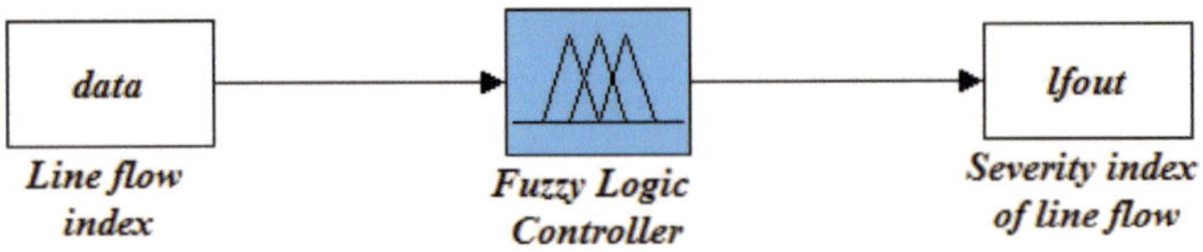

Figure 13.
Line flow fuzzy logic controller block.

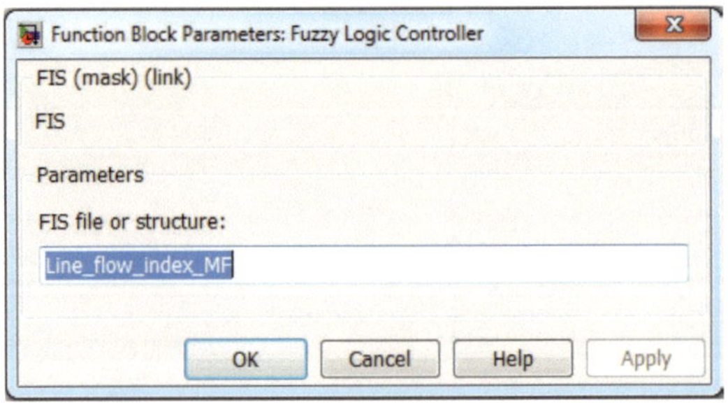

Figure 14.
Inserting of voltage profile MF on fuzzy logic control block.

severity index for the specific contingency case. The CI can be identified mathematically as the common summation of total severity indices of voltage profile ($\sum$SIVP), and total severity indices of line flow ($\sum$SILF), where both ($\sum$SIVP), and ($\sum$SILF) are the outputs of the two blocks in **Figures 11** and **13**, respectively.

5. Fuzzy logic simulation results analysis compared to NEPLAN contingency ranking

In this section, the results of ranking through the fuzzy logic control system have been compared with the ranking that had previously obtained from the NAPLAN program. The first subsection represents the NAPLAN results as a reference. The second subsection represents the detailed results from fuzzy interface system (FIS) blocks. The results are obtained for all elements in the implemented network but for simplification purposes, only the most severe five contingencies had been reviewed.

5.1 Results of single contingency ranking from NEPLAN software

Table 2 represents the NEPLAN ranking for the most severe five contingencies that had been used as reference ranking.

Contingency No.	Type of contingency	From	To	Violation (%)	Violated element	Ranking
1	Line 23	Bus 61	Bus 62	142.63	Line 36	1
2	Line 37	Bus 66	Bus 61	110.72	Node 66	2
3	Line 20	Bus 53	Bus 54	88.52	Node 61	3
4	Line 26	Bus 59	Bus 60	88.17	Node 73	4
5	Line 10	Bus 1	Bus 2	83.82	Node 12	5

Table 2.
Selected contingency from the contingency results of NEPLAN.

5.2 Results analysis of fuzzy logic system contingency ranking

This subsection represents the fuzzy logic control ranking results for the same most severe single contingencies which had been illustrated in Section 5.1.

5.2.1 Contingency1: outage of line 23

Table 3 represents the severity indices for each node's voltage profile when an outage on line 23 occurred.

Node name	Voltage profile (p.u)	SI_{VP}
61	0.9240	43.75
21	0.9245	43.75
63	0.9272	43.75
62	0.9413	43.75
54	0.9507	43.75
53	0.9509	43.75
58	1.0443	43.75
55	0.9572	43.2685
26	0.9575	43.2042
56	0.9591	42.2508
25	0.9595	41.3088
58	0.9609	40.6618
11	1.0384	40.6618
57	0.9624	40.6618

Node name	Voltage profile (p.u)	SI_{VP}
64	1.0371	40.6618
65	1.0339	39.7147
48	0.9665	39.6324
52	0.9665	39.6324
10	1.0315	39.6324
67	1.0307	39.6324
13	0.9694	37.9946
66	1.0306	37.6856
29	0.972	37.6856
9	1.0256	37.6856
51	0.9764	37.6856
7	1.0216	35.5912
6	1.0204	35.5912
69	1.0203	35.5912
68	1.0199	35.5912
3	0.9807	35.1786
4	0.983	34.8444
73	1.017	34.549
47	0.9835	34.549
22	0.9844	34.549
23	0.9861	34.549
50	0.9861	34.549
70	1.0126	34.549
15	0.9879	34.1867
16	0.9884	34.121
5	0.9892	33.8188
59	0.9896	33.445
30	0.9900	33.0653
14	1.0087	32.6796
43	0.9916	32.2876
20	1.0083	31.8893
42	0.9918	31.4844
60	1.0081	30.9167
32	1.0079	30.9167
36	1.0072	26.25
72	1.0062	26.25
24	0.9941	26.25
35	1.0054	26.25
40	1.0046	26.25
2	1.0039	26.25
45	1.0038	26.25
12	0.9964	26.25

Node name	Voltage profile (p.u)	SI_{VP}
19	0.9971	26.25
27	0.9974	26.25
39	1.0026	26.25
44	0.9975	26.25
17	0.9978	26.25
37	1.0019	26.25
41	0.9986	26.25
46	0.9986	26.25
34	0.9987	26.25
33	0.9988	26.25
38	1.0006	26.25
1	1.00	26.25
18	1.00	26.25
28	1.00	26.25
31	1.00	26.25
49	1.00	26.25
71	1.00	26.25
$\sum SI_{VP}$		**2539.404**

**The yellow line represents the buses (61,62) which link line 23 where both busses the most higher SI_{VP} in the table and also the results represent that line 23 has the highest line flow severity index [11].*
Buses (59,68) in **Table 3 link the transmission line in parallel with line 26 in contingency no4 this means that the outage of this line will have a severity index near equal to line 26 [11].*

Table 3.
Severity indices for line 23 contingency [11].

Line 23 takes into consideration the effect of the high violation on line 36, which reached (142.63%) and the nodes with the contingency of line 23 connected between them are node (61) and node (62), both two nodes have a severity index of (43.75) which classified as the highest value for the severity indices on **Table 3** and defined as More Severe (MS) on the membership function editor at the fuzzy set. On the other hand, the violated element itself connected through the nodes (59) and (68); their severity indices are (33.4450 and 35.5912), respectively, and this is defined as below severe (BS) as set in **Table 1** in Section 4.1.

As shown in **Table 4**, the light concentrated on the line flow index for line 36 which is the violated element for this contingency of line 23 outage. The severity index is (13.7797) which is defined on the fuzzy set as the less severe (LS) range.

When these results are compared with NAPLAN results, line 36 has a violation of (89.50%), that is, less than 100% which is below the severity limit of the network. From the whole mentioned above, it can be judged that by the contingency of line 36, the system is secure.

By returning to the main contingency of line 23, the composite index (CI) must be accounted for to rank the contingency of line 23. From the defuzzification results in **Tables 3** and **4**:

$$CI_{\text{line}23} = \Sigma(\Sigma \text{SIVP} + \Sigma \text{SILF})_{\text{line}23} = 2539.4040 + 429.2228 = \mathbf{2968.6268} \quad (1)$$

Line name	L.F index	SI_{LF}
36	0.8645	13.7797
25	0.8589	13.0675
24	0.6227	12.3651
26	0.5798	12.1934
35	0.5736	12.1934
32	0.5637	11.5047
31	0.5637	10.816
34	0.5619	10.3066
22	0.5612	10.3066
30	0.5581	9.6066
20	0.5425	8.8968
22	0.5391	8.3654
17	0.4806	8.3654
16	0.4783	7.9602
18	0.4169	7.7078
14	0.4161	7.6124
15	0.375	7.1147
13	0.3693	6.8377
51	0.3622	6.4728
44	0.3598	6.25
41	0.3589	6.25
78	0.3578	6.25
43	0.3433	6.25
38	0.3408	6.25
53	0.333	6.25
52	0.3122	6.25
1	0.2714	6.25
3	0.2713	6.25
37	0.2683	6.25
74	0.2638	6.25
39	0.2621	6.25
19	0.262	6.25
10	0.2604	6.25
4	0.2568	6.25
75	0.2565	6.25
76	0.2565	6.25
2	0.2552	6.25
71	0.2541	6.25

Line name	L.F index	SI_{LF}
49	0.2444	6.25
42	0.2398	6.25
41	0.2371	6.25
45	0.2371	6.25
61	0.2367	6.25
63	0.2365	6.25
56	0.2364	6.25
59	0.2236	6.25
57	0.2234	6.25
69	0.2202	6.25
67	0.2193	6.25
50	0.2192	6.25
5	0.2186	6.25
68	0.2076	6.25
62	0.1965	6.25
64	0.1947	6.25
60	0.1866	6.25
65	0.183	6.25
70	0.1823	6.25
6	0.1801	6.25
$\Sigma SI_{L.F}$		**429.2228**

**The yellow line represents line 30 with the highest SILF indicating that it is the most severe line in contingency1 [11].*
The yellow line in line 36 results represents to its severity ranking according to its link with bus 59 in **Table 3 [11].*

Table 4.
Severity indices for line flow index [11].

The previous illustration can be represented clearly by graphs in **Figures 15** and **16**.

5.2.2 Contingency2: outage of line 37

The results of a single contingency of line 37 can be represented and clarified as in **Figures 17** and **18**.

Figure 17 represents the severity indices voltage profile against the voltage profile when a single contingency of line 23 occurs on the system.

From the network in **Figure 1**, transmission line 37 is connected through basbars 66 and 61. Hence, the severity index for these two nodes represents the highest value, for instance, node 66 violated with (110.7%), and the severity indices for node 61 and its around nodes is (43.75) which is located in more severe (MS) at the fuzzy set.

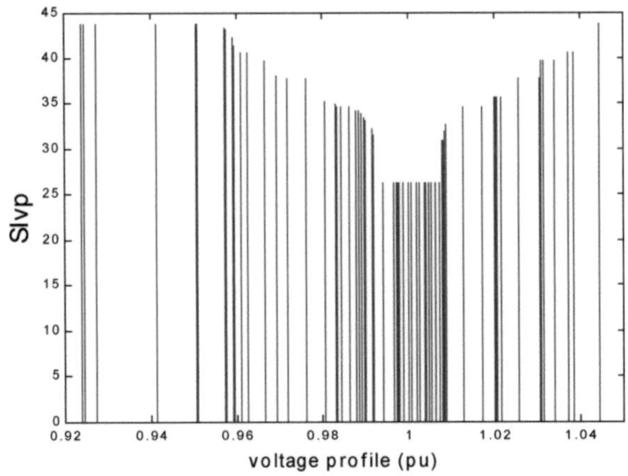

Figure 15.
Severity index of voltage profile for the contingency of line 23.

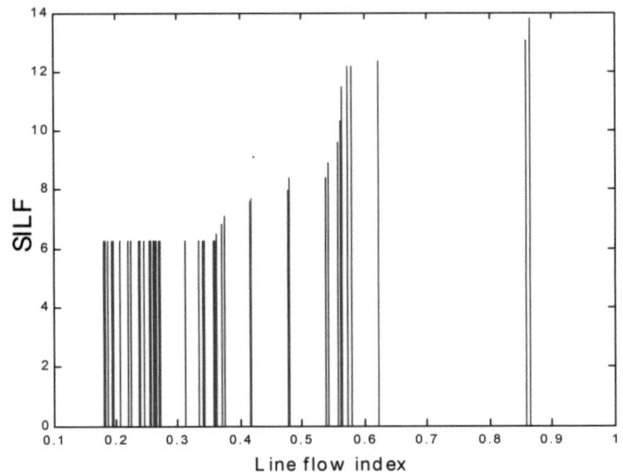

Figure 16.
Severity index of line flow for the contingency of line 23.

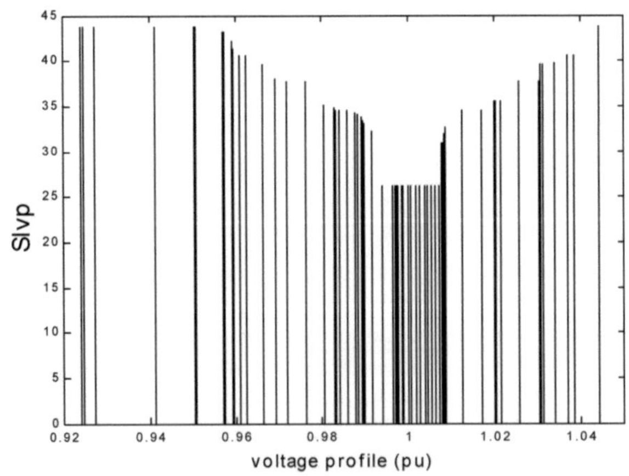

Figure 17.
Severity index of voltage profile for the contingency of line 23.

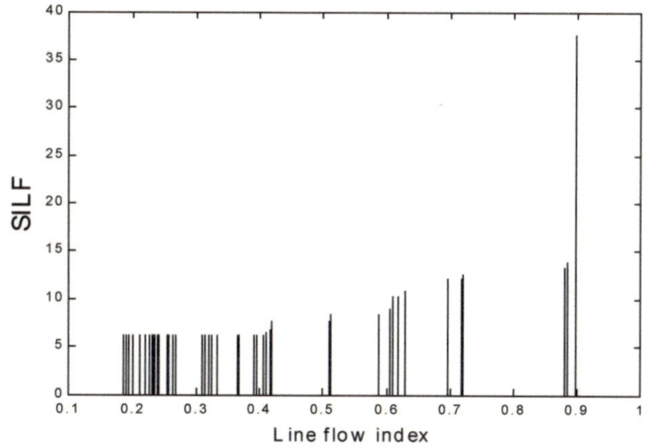

Figure 18.
Severity index of line flow for the contingency of line 37.

From the results of line flow severity indices shown in **Figure 18**, it is clear that the contingency of line 23, has a severity index of 37.623 which is defined in a range of above severe (AS) at a fuzzy set of membership function, although the other lines have a severity index classified as (LS) and (VLS) at fuzzy logic setting as in **Table 2**, Section 4.1.

With the typical defuzzification process, the results of various severity indices for line 37 contingency can be obtained in tables as in contingency 1. It found that the value of ($\sum$SIVP) is **2494.3840**, and the value of $\sum$SILF is **428.9414**. Hence, the composite index ($CI_{line\ 37}$) can be obtained as:

$$CI_{line37} = \Sigma(\Sigma SIVP + \Sigma SILF)_{line37} = 2494.3840 + 428.9414 = \mathbf{2923.3254} \quad (2)$$

5.2.3 Contingency3: outage of line 20

The defuzzification results of the contingency line 20 shown in **Figures 19** and **20**.

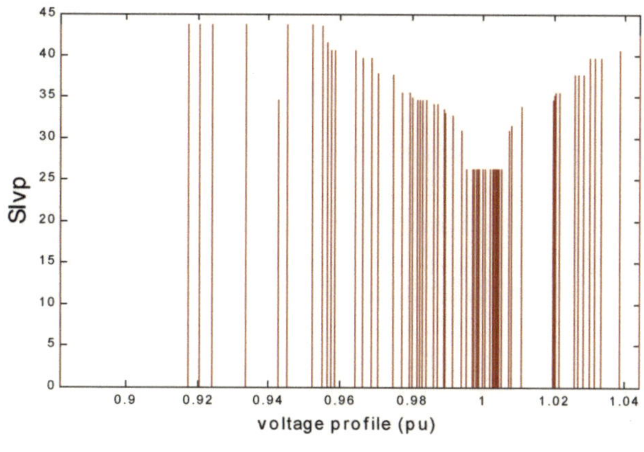

Figure 19.
Severity index of voltage profile for the contingency of line 20.

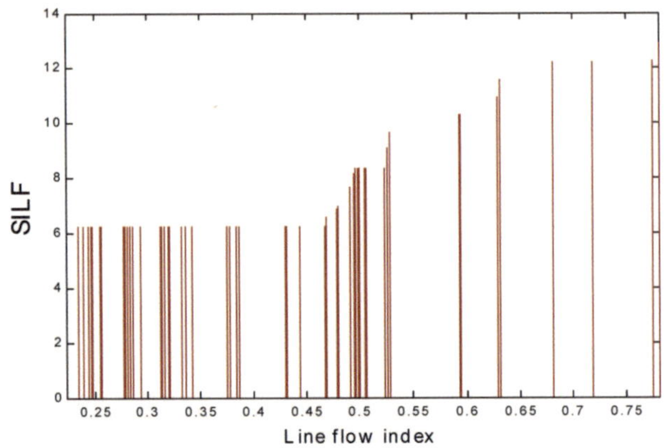

Figure 20.
Severity index of line flow for the contingency of line 20.

Figure 19 represents that the highest (SIVP) value is 43.75, which appeared at nodes 53 and 54, respectively, because they are the holder nodes for line 20 itself. The severity spread around its neighbor directly connected nodes that had possessed the same severity index. The remaining nodes on the network have severity indices in the range of (MS) set.

Regarding the transmission lines flow severity indices shown in **Figure 20**, most SILFs are in the range of very less severe (VLS) and limited lines in less severe (LS) according to the fuzzy set in **Table 1**, Section 4.1. An outage of line 20 has a high influence on system voltage stability that most of the nodes had an (MS) severity index; this can be noticed clearly from Eq. (3) that the weight of the composite index consists of high ($\sum$SIVP). The composite index of line 20 (CIline 20) contingency is:

$$CI_{\text{line}\,20} = \Sigma(\Sigma\text{SIVP} + \Sigma\text{SILF})_{\text{line}\,20} = 2461.7530 + 451.8530 = \mathbf{2913.6060} \quad (3)$$

Figures 19 and **20** show the severity indices of line 26 outage.

5.2.4 Contingency4: outage of line 26

The single contingency of line 26 will be illustrated in this section with the aid of **Figures 21** and **22**.

Figure 21 represents SIVP on line 26 contingency against the voltage profile on the line; the most severe element is node 73, and its connected nodes are affected by it with high severity index voltage profile (SIVP) and the remaining nodes at the network had been possessed (SIVP) in the (BS) limits at the fuzzy set.

Figure 22 represents the values of SILF according to the line flow index in the network at the outage of line 26. Most of the severity indices of line flow (SILF) for this contingency are within limits of (VLS), and fewer of them are within (LS) limit as in the fuzzy set in **Table 1**. As a result, it can be noticed that the contingency of line (26) outage had a less ($\sum$SILF) impact on the composite index, this can be shown in Eq. (4):

$$CI_{\text{line}\,26} = \Sigma(\Sigma\text{SIVP} + \Sigma\text{SILF})_{\text{line}\,26} = 1708.767 + 344.5316 = \mathbf{2053.2986} \quad (4)$$

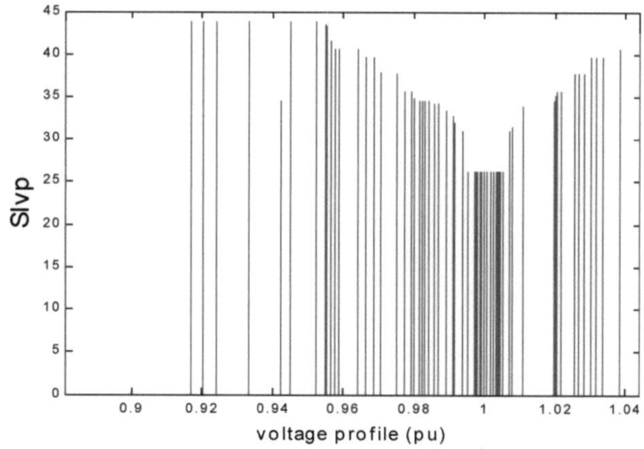

Figure 21.
Severity index of voltage profile for the contingency of line 26.

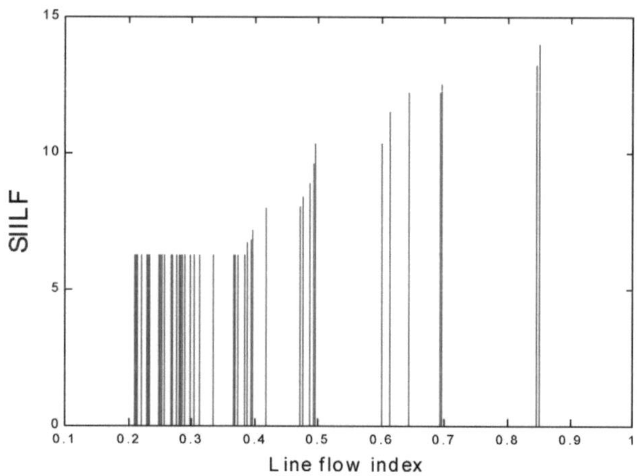

Figure 22.
Severity index of line flow for the contingency of line 26.

5.2.5 Contingency5: outage of line 10

In this subsection, results for line 10 single contingency, will be clarified as shown in **Figures 23** and **24**.

Figure 23 represents the magnitudes of SIVP depending on the voltage profile at the outage of line 10. In this contingency, the most affected node is node 12 that it violated with (83.82%), the other bus-bars on the network had possessed severity indices of voltage profile (SIVP) in the limits of below severe (BS) on the fuzzy logic classified previously in Section 4.1. Hence, it is worth mentioning that the single contingency of line 10 had a less severe influence on the voltage security.

Figure 24 shows the SILF according to the implemented elements load flow index, when returning to the network diagram shown in **Figure 1** in Section 3, transmission line 10 is connected through busbar 1 and busbar 2, so, it is normally to find that the most affected lines with line 10 contingency, those which closely connected to

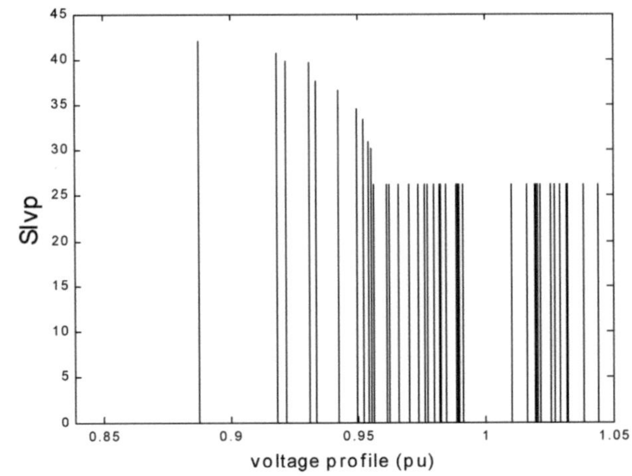

Figure 23.
Severity index of voltage profile for the contingency of line 10.

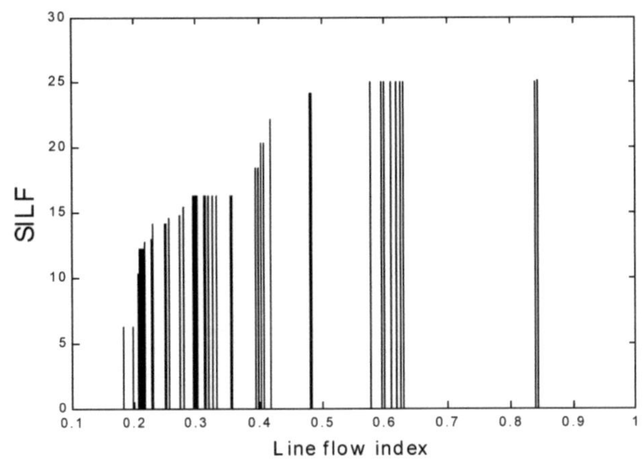

Figure 24.
Severity index of line flow for the contingency of line 10.

busbar 1 and busbar 2 that they had been in the top of severity indices list in the limits of below severe as a fuzzy setting in **Table 1**. The remaining transmission lines of the system possessed values SILF in limits between (LS) and (VLS) of severity index for the other lines are varied within the range of (LS) and (VLS).

The range of total algebraic summation for the severity indices voltage profile in this contingency is less value compared with that for severity indices line flow which had been arranged in limits of (BS, LS, and VLS).

According to which mentioned above, it can be handed down that line 10 single contingency has a high impact on system power loading (MW) which can give the operator engineer an indicator to concentrate on the megawatt administration techniques to monitor this contingency. Eq. (5) below represents the composite index for line 10 contingency (CIline 10) is:

$$CI_{\text{line10}} = \Sigma(\Sigma \text{SIVP} + \Sigma \text{SILF})_{\text{line10}} = 812.3472 + 987.7966 = 1800.1438 \quad (5)$$

Cont. No.	Type of contingency	Violation (%) from NEPLAN	Ranking of contingency through NEPLAN	CI $\sum(\sum SI_{VP}, \sum SI_{LF})$ from fuzzy interface system	Ranking of contingency through fuzzy logic approach
1	Line 23	142.63	1	2968.6268	1
2	Line 37	110.72	2	2923.3254	2
3	Line 20	88.52	3	2913.6060	3
4	Line 26	88.17	4	2053.2986	4
5	Line 10	83.82	5	1800.1438	5

Table 5.
Compression of fuzzy logic system contingencies ranking with the NEPLAN contingency ranking for the same five single contingencies.

5.3 Results compression of contingencies ranking

Table 5 shows that any element of the selected contingencies has the same ranking within both the NAPLAN program and the severity indices obtained from the fuzzy logic system (FIS) [18].

6. Conclusion

This chapter represented an intelligent technique based on the fuzzy logic controller for ranking a single contingency. The fuzzy logic ranking depends on the severity index (SI) denoted with the composite index (CI) which is accumulated of two severity indices, the first one, is the severity index of voltage profile (SIVP) and the other is the severity index of line flow (SILF).

In this chapter, the technique of fuzzy logic Interface system has been applied to a partial Sudanese transmission network. The results were compared to those obtained from NEPLAN software and the fuzzy logic system approach had been approved for the same ranking with the additional advantage that it is more reliable and robust in ranking the single contingencies. Also, contingency ranking based on CI imparted the fuzzy logic technique more precession because it indicates the source of insecurity on the network that the security problem refers to voltage problem if ($\sum$SIVP) is higher than ($\sum$SILF) on the (CI), and the security problem refers to power flow on the network if the ($\sum$SILF) is the higher component of CI, this significant issue because it guides the network dispatcher to take the proper control action according to system deficiency.

7. Recommendations

- Applying the fuzzy logic ranking technique on the extended actual grids for its high reliability.
- Upgrading the technique by linking it through the adaptive neuro-fuzzy interface system (ANFIS) function in MATLAB and simulating it through the Sugeno-model system instead of the Mamdani model which is used in this chapter.

Author details

Israa Salih H. Alajab[1*] and Mohammed Osman Hassan[2]

1 Sudanese Electricity Transmission Co. Ltd, Khartoum, Sudan

2 Sudan University of Science and Technology, Khartoum, Sudan

*Address all correspondence to: israamm9119@gmail.com

References

[1] Mikklii S, Panda AK. Real-time implementation of PI and fuzzy logic controllers based shunt active filter control strategies for power quality improvement. International Journal of Electrical Power & Energy Systems. 2012;**143**(1):1114-1126. DOI: 10.1016/j.ijepes.2012.06.045

[2] Iancu I. A Mamdani Type Fuzzy Logic Controller. Intech Open, Open Access books. ISBN: 978-953-51-0396-7; 2012

[3] Lo KL, Abdelaal AKI, et al. Fuzzy logic based contingency analysis. In: IEEE Conference; 04-07 April 2000. London UK: IEEE; 2000. DOI: 10.1109/DRPT.2000.855715

[4] Krishnakumar B, Subaashini M, Ganesh Kumar E, Arthi R. Power flow based contingency analysis using fuzzy logic. Procedia Engineering. 2012;**38**:3603-3613

[5] Manjula BG, Suma AP, Kulkarni AD, Kumar N, Ananthapadmanabha T. Contingency ranking in power systems employing fuzzy based analysis. International Journal of Electrical and Electronics Engineering Studies (IJEEES). 2012;**4**(1):1-7

[6] Alam M. Determination of Power System Contingency Ranking Using Novel Indices. Singapore: Springer Nature; 2022

[7] Nagendra Prasad HK, Kantharaj B, Kavitha KM, Ramesh E. Contingency analysis of a power system by using fuzzy approach. International Journal of Engineering Research & Technology (IJERT). 2014;**3**(6):1659-1661. ISSN: 2278-0181

[8] Mohsena S, Seyavashc C, Sepided Y, Alie S. A new method for power system contingency ranking using a combination of neural network and data envelopment analysis. Journal of Intelligent & Fuzzy Systems. 2017;**32**(6):3859-3866. DOI: 10.3233/IFS-162169

[9] Ramesh E, Prakash R, Lekshmi M, Yogeesh S. Case study on fuzzy logic based network contingency ranking. International Journal of Engineering Research & Technology (IJERT). 2013;**2**(7):830-834. DOI: 10.17577/IJERTV2IS70268

[10] Hassan M, O, Alajab ISH. Sudan national grid contingency ranking through fuzzy logic approach. International Journal of Advanced Engineering and Nano Technology (IJAENT). 2022;**9**(4):3-6. DOI: 10.35940/ijaent.C0468.049422. ISSN: 2347-6389 (Online)

[11] Alajab ISH, Hassan MO. Thesis of: "National Sudanese Grid Contingency Ranking through Fuzzy Logic Approach". Sudan: Sudan University; 2017. Available from: http://repository.sustech.edu/handle/123456789/18488

[12] Poshtan M, Rastgoufard P, Singh B. Contingency ranking for voltage stability analysis of large scale power systems. In: Proceeding of IEEE/PES Power System Conference and Exposition. 2004. pp. 1595-1602

[13] Mohammed MM. Development of Intelligent Systems for Evaluating Voltage Profile and Collapse under Contingency Operation. Egypt: M. S., Zagazig University; 2004

[14] Shankar S, Ananthapadmanabha T. Fuzzy Approach to Critical Bus Ranking Under Normal and Line Outage Contingencies. Dubai, UAE:

International Journal of Soft Computing (IJSC); 2011;**2**(1)

[15] Magaji UA. Contingency analysis using fuzzy logic. In: 2nd International Conference on Science, Technology and Management, University of Delhi (DU), Conference Centre, India; 27 September, 2015

[16] Fuzzy Logic Toolbox User's Guide "Fuzzy Logic Toolbox for Use with MATLAB", COPYRIGHT by The MathWorks, U.S; 1995-1999. pp. 6-63

[17] Christy, Priscilla D, Christy A. Contingency ranking method using fuzzy controller. International Journal of Innovative Technology and Exploring Engineering (IJITEE). (Online). 2019;**8**(12):3104-3108. ISSN: 2278-3075

[18] Zanzoni A, Montecchi-Palazzi L, Quondam MX. Mint: A molecular interaction database. FEBS Letters. 2002;**513**:135-140. DOI: 10.1016/s0014-5793(01)03293-8

Chapter 2

The Design of Fractional-Order Robust Adaptive Intelligent Controller for Fractional-Order Nonlinear Systems with Time Delay, Uncertainty and Disturbance

Mehdi Dalir and Nooshin Bigdeli

Abstract

Recently, many practical systems have been described via nonlinear fractional-order dynamic models. The design of controller for such systems is not a straightforward task, where their complex behaviors may result in chaos and bifurcation. This challenge becomes more serious when the system experiences time delays, uncertainties and disturbances. This chapter presents the controller design strategies for fractional-order nonlinear systems with time delays, uncertainties and disturbances via robust adaptive intelligent structure. Time-varying heterogeneous time delays are considered in inputs and states; and two robust adaptive intelligent strategies are designed and compared based on sliding mode and backstepping control methods. Besides, the stability analysis of the controllers is carried out based on Lyapunov-Krasovskii method and Barbalat's Lemma. Simulation examples show the effectiveness of the proposed method for such systems. To demonstrate the effectiveness of Adaptive Network-based Fuzzy Inference System (ANFIS) on designed controller, their performance is compared with that of fractional-order robust adaptive controllers, as well.

Keywords: adaptive intelligent control, neuro-fuzzy estimator, time delay, robust control, fractional-order nonlinear system, uncertain fractional-order system

1. Introduction

Due to their good capabilities in modeling complicated systems, fractional-order nonlinear dynamics have been widely used for describing complex natural phenomena as well as real engineering, physical and biological systems. Such systems perform complicated behaviors such as bifurcation and chaos and so, the design of controllers for fractional-order nonlinear systems is not a straightforward task. These challenges become more serious when the system experiences time delays, uncertainties and disturbances, as well. Time delays may be observed in many systems, including chemical processes (chemical behavior kinetics), biological sciences (inheritance in population dynamics), mechanical systems (hydraulic-pneumatic networks), economics (business cycle), telecommunication systems, power systems, robot control, etc. [1–5]. Time delay may occur in the input, output, states or feedback loop of a system. When describing real systems, the delays may also be time-varying. Although robust adaptive controllers may improve the closed-loop behavior of the disturbed uncertain nonlinear systems, the existence of time delay in such a system may degrade the performance of the controllers, drastically. In addition, time delay may imply chaos in the nonlinear system. Therefore, stabilizing delayed nonlinear fractional-order systems is more difficult compared to non-delayed systems. In this context, seeking for an efficient infrastructure for stabilizing and performance control of fractional-order nonlinear systems with time delays, uncertainties and disturbances seems important.

In the literature, robust controllers in conjunction with adaptive strategy have been widely used for chaos control of uncertain nonlinear systems. In this context, sliding model control and backstepping control as well as hybrid controllers are employed to achieve finite-time stability, fast response, good transient response performance and robustness with respect to disturbances and uncertainties. These controllers could provide asymptotic stability in systems with severe uncertainties and disturbances to some extent. Addition of adaptive property to the controllers resulted in adaptive robust controllers that improved the closed-loop performance of the system by estimating unknown system parameters and dynamics over a wide range of working conditions including external disturbances and unmodeled dynamics. These controllers were next improved by employing hybrids of conventional controllers along with machine learning-based control infrastructures, including neuro-fuzzy estimators. Therefore, the so-called adaptive intelligent robust control infrastructure consists of adaptive neuro-fuzzy approximators serving as the system approximators and a (hybrid) robust controller. This infrastructure outperforms the adaptive robust controllers, considerably due to its good system approximation capability. However, introduction of time delays to these systems makes the controller design more challenging. This is the case especially for uncertain nonlinear fractional-order systems with heterogeneous time-varying delays. In this area, not only are there considerably less researches dealing with this topic, but there is also no comprehensive resource to integrate the researches altogether.

Based on the provided discussion, this chapter presents the controller design strategies for fractional-order nonlinear systems with time delays, uncertainties and disturbances via robust adaptive intelligent infrastructure. Time-varying heterogeneous time delays are considered in inputs and states; and two robust adaptive intelligent strategies are designed and compared based on sliding mode and backstepping control methods. Besides, the stability analysis of the controllers is carried out based on Lyapunov-Krasovskii method and Barbalat's Lemma. Simulation examples show the effectiveness of the proposed method for such systems. To demonstrate the effectiveness of ANFIS on designed controller, their performance is compared with that of fractional-order robust adaptive controllers, as well.

1.1 Related works

As stated, in the recent years, the use of hybrids of intelligent and classic controllers to control nonlinear systems has been highly regarded. In Sun et al. [6] in order to control strict-feedback nonlinear systems, a finite-time adaptive intelligent robust controller has been designed. An adaptive fuzzy control has also been employed in Sun et al. and He et al. [7, 8] to control switched nonlinear systems with external disturbances. In these articles, adaptive fuzzy control is used to estimate unknown nonlinear functions, and a backstepping control is also proposed as robust control of switched multi-input and multi-output (MIMO) nonlinear system. In Wang et al. [9], an adaptive intelligent robust controller has been presented in order to track the target of the autonomous underwater vehicles (AUVs). Fuzzy logic is also used in tracking control to estimate the dynamic uncertainties of the system. Intelligent adaptive control design for multi-agent systems with unmodeled dynamics, external disturbances and dead-zone input has been done in Liang et al. and Zhou and Tong [10, 11]. In the proposed controller, neural networks of radial basis function are used to approximate the unknown nonlinear function and unmodeled dynamics. Fuzzy adaptive backstepping control has been designed to control strict-feedback nonlinear system with input delays in Tong et al. [12]. A designed observer based on fuzzy logic has been used to estimate the unknown functions. Also, robust stability of the studied system is guaranteed by using output-feedback backstepping control. In Huang et al. [13], the hybrid of neural network control with adaptive feedback control has been used to control a class of nonlinear systems. The proposed controller included an adaptive backstepping control section. The nonlinearity and unknown dynamics of the system were also estimated using the radial-basis-function neural networks.

Fractional-order differential equations serve as a proper tool for describing different dynamic systems. Due to the wide applications of fractional-order nonlinear models, the controller design for these systems has become an important issue. So that the robust adaptive intelligent control has been presented in Bahraini et al. and Liu et al. [14, 15] to control the nonlinear fractional-order systems. The uncertainty of the system has also been included in the design of the controller. The proposed controller in Bahraini et al. [14] included a supervised sliding mode control and a fractional-order intelligent adaptive Proportional integral Derivative (PID) controller whose coefficients are updated using fuzzy rules. The use of adaptive neural network control to estimate uncertain parameters and dynamics is presented in Liu et al. [15]. It has also ensured the robust stability of the system by using the feedback controller. The design of the adaptive fuzzy controller has been done for the fractional-order nonlinear large-scale systems in Zhan et al. and Zhan and Tong [16, 17] as well as the fractional-order nonlinear switched systems in Bi et al. [18]. In these articles, the adaptive fuzzy control strategy has been used in order to estimate uncertainties. The hybrid of adaptive control and neural network has also been used to control the nonlinear fractional-order system in Bi [19], Song et al. [20], and Zouari et al. [19–21]. Neural network was used to estimate unmodeled dynamics and uncertainties. Also, in order to avoid the "explosion of complexity" issue, a backstepping control was designed. In You et al. and Ma and Sun [22, 23], an adaptive fuzzy controller was used for the nonlinear fractional-order system, despite the external disturbance. Based on the features of the fuzzy system as well as adaptive feedback controllers, a robust adaptive intelligent control was presented to control the mentioned system.

Time delay is also an inseparable part of system modeling. The presence of time delay in the design of the controller itself is a challenging issue that has been the focus of many authors in recent years. In Song et al. [24], a combination of fuzzy controller with adaptive controller and output feedback controller was used to control a class of fractional-order nonlinear systems with time-varying delay in system states. The design of adaptive fuzzy controller for fractional-order nonlinear system with time delay was given in Dong et al. and Montazeri et al. [25, 26]. In Wang et al. [27], a neural adaptive control has been considered for the fractional-order nonlinear system with state delay and uncertainty. The design of a fuzzy adaptive controller for nonlinear multi-agent system with time delay has been given in Li et al. [28] and for fractional-order networked systems in Yang and Xhang [29]. Stability analysis of fractional-order delay systems using Takagi-Sugeno (TS)-Fuzzy method has also been discussed in Liu et al. and Yang et al. [30, 31]. Yifei Xing et al. in Xing and Wang [32] used the hybrid of backstepping adaptive controller with intelligent control in order to control the nonlinear fractional-order system with input delay. Also, in order to control the multi-agent fractional-order system with time-varying delay, a sliding mode intelligent controller was proposed in Yan et al. [33]. In Zhang et al. and Liu et al. [30, 34], a fuzzy robust controller for fractional-order nonlinear system with time delay was presented. Fuzzy logic is used as an estimator in these articles. In Zhang et al. [34] by using the spatial linear matrix inequality method and also in Parvizian et al. [35], an adaptive sliding mode controller based on H_∞ guaranteed the robustness of the controller.

In summary and as mentioned above, the works in this area can be categorized to three categories: Those works mentioned in Sun et al. [6], Sun et al. [7], He et al. [8], Wang et al. [9], Liang et al. [10], Zhou and Tong [11], Tong et al. [12] and Huang et al. [6–13] concentrate on the hybrid of the classic and intelligent controllers for control of integer-order nonlinear systems with uncertainty and disturbance. The focus in Bahraini et al. [14], Liu et al. [15], Zhan et al., [16], Zhan and Tong [17], Bi et al. [18], Bi [19], Song et al. [20], Zouari et al. [21], You et al. [22], Ma and Sun [14–23] was the design of fractional-order hybrid controllers for fractional-order nonlinear systems with uncertainty and disturbance. Next, in Song et al. [24], Dong et al. [25], Montazeri et al. [26], Wang et al. [27], Li et al. [28], Yang and Zhang [29], Liu et al. [30], Yang et al. [31], Xing et al. [32], Yan et al. [33], Zhang et al. [34], Parvizian et al. [24–35], one may find that the design of these controllers for fractional-order nonlinear systems with time delay, uncertainty and external disturbance was investigated.

In this chapter, the design of an adaptive intelligent resilient control will be discussed to control the fractional-order nonlinear system with time delay, uncertainty and disturbance. Unlike the recently reviewed works, in a more comprehensive manner, the time-varying delay in the states, the control input and the uncertainty of the system are considered. The controller introduced here is based on tangential hyperbolic sliding mode control strategy, which reduces the response time and increases the accuracy of the controller.

2. Problem statement

As stated earlier, in this chapter, a fractional-order adaptive intelligent robust control will be presented for controlling fractional-order nonlinear systems with uncertainty, disturbance and time delay. The time delay is assumed to be generally time varying, and in addition to the system states, it exists in the control input as well as the uncertainty of the system. It is expected that the use of fractional-order adaptive intelligent robust controller not only stabilizes the system, but also reduces fluctuations, and makes the system respond faster and more accurately.

In general, the model of a delayed uncertain fractional-order nonlinear system subject to external disturbances is presented by Eq. (1):

$$\begin{cases} D^q x(t) = f(x) + \Delta f(x) + g(x(t-\tau)) + \Delta g(x(t-\tau)) + d(t) + u(t-\tau_u) \\ x(t) = p(t), \quad t \in [-\tau, 0] \\ u(t) = q(t) \quad t \in [-\tau_u, 0] \end{cases} \tag{1}$$

Where, $x(t) \in R^n$ is the state vector, $u(t) \epsilon R^m$ is the control input vector, $f(x)$ and $g(x(t-\tau))$ are unknown continuous functions, $\Delta f(x)$ and $\Delta g(x(t-\tau))$ are the system uncertainties, $d(t)$ is external disturbance, $p(t)$ and $q(t)$ are also initial conditions of the states and inputs, respectively. Besides, τ_u is input time delay and τ is also the state time delay considered as time-varying delays and satisfying:

$$t - \tau > 0 \tag{2}$$

$D^q \in R^n$ is also the Caputo fractional derivative operator of order q. For a given continuous function $f : R^+ \to R$, Caputo fractional-order derivative is defined as:

$$D^q f(t) = \begin{cases} \dfrac{\int_0^t \dfrac{f^{(l)}(\tau)}{(t-\tau)^{q-l+1}} d\tau}{\Gamma(l-q)}, l-1 < \quad q < l \\ \dfrac{d^l}{dt^l} f(t), \quad q = l \end{cases} \tag{3}$$

Where Γ is the well-known Gamma function, and $\Gamma(z) = \int_0^t e^{-t} t^{z-1} dt$, and $\Gamma(z+1) = z\Gamma(z)$.

To continue, the design of the controller for system (1) is discussed. At first, the desired controller is designed for the system with a time delay in states and then the controller is expanded for the system (1) with a time delay in states and inputs. For this purpose, at first, Eq. (1) is simplified and rewritten in the following form as:

$$D^q x(t) = f(x) + g(x(t-\tau)) + l(t,x) + u(t) \tag{4}$$

Where, system uncertainties are lumped as $l(t,x) = \Delta g(x(t-\tau)) + \Delta f(x) + d(t)$. Next, two types of fractional-order adaptive intelligent robust controllers based on sliding mode control and backstepping control will be designed. To continue, initially the general structure of these control strategies is described and next, the controller design procedure is described step by step.

3. The structure of fractional-order robust adaptive intelligent controllers

Figure 1 shows a general structure of the closed-loop control system when the adaptive intelligent robust control strategy is employed. As seen, in this infrastructure, sliding mode control and backstepping control methods are employed as robust controllers. In the design of the robust controller, tangential hyperbolic sliding surfaces are used due to its capability to avoid high tracking errors while it achieves fast response, good transient performance and robust stability with respect to disturbances and uncertainties. It is assumed that the system states are available and the sliding surfaces are calculated by the fed back state values. On the other hand, for approximating model uncertainties and dynamics, an adaptive neuro-fuzzy inference system (ANFIS) is employed as an intelligent estimator. In addition, the adaptive

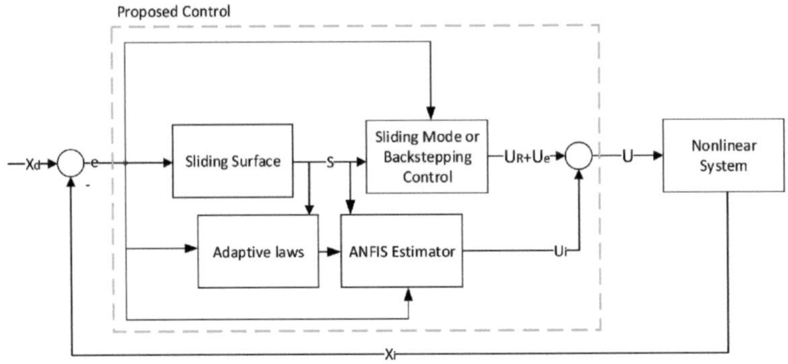

Figure 1.
The block diagram of closed-loop control method.

parameters of the control system are presented online by adaptive laws using the model reference method.

3.1 The adaptive neuro-fuzzy inference system (ANFIS) as system model approximator

IF-THEN fuzzy rules are used as an ANFIS and approximator in designed control. The design basis of this estimator is described as follows [36]:

$$Rule - j:$$

$$IF\ (x_1\ is\ A_{1j}, x_2\ is\ A_{2j}, \dots, and\ x_p\ is\ A_{pj})\ THEN\ \hat{F}_j = \sum_{k=1}^{M} w_{kj} y_k = w_{1j} y_1 + w_{2j} y_2 + \dots + w_{Mj} y_M \quad (5)$$

Where, ANFIS input and local output are denoted by x_i and $\hat{F}_j$, respectively. p is the number of input variables; A_{ij} expresses the linguistic label of the preconditioned part with Gaussian membership function; $w = [w_{1j}, w_{2j}, \dots, w_{Mj}]$ is the vector of weights; y_k and M indicate the basis function and number of them, respectively.

In other words, as can be seen in **Figure 2**, the ANFIS general structure contains five layers. The steps of the estimator are as follows:

Layer 1 (input layer): Each node in this layer is considered as an input.

$$y_i^{(1)} = x_i, i = 1, 2, \dots, p \quad (6)$$

Layer 2 (Membership function layer): The membership function value is calculated in this layer. Also, a single language label corresponding to the first-layer input variable corresponds to each node. Therefore, the membership degree of an input to the fuzzy set in Layer 2 is determined by the membership value. The Gaussian function related to this layer is defined as follows:

$$y_{ij}^{(2)} = exp\left(-\frac{\left[y_i^{(1)} - m_{ij}\right]^2}{2\,\sigma_{ij}^{2}}\right), i = 1, 2, \dots, p; j = 1, 2, \dots, M \quad (7)$$

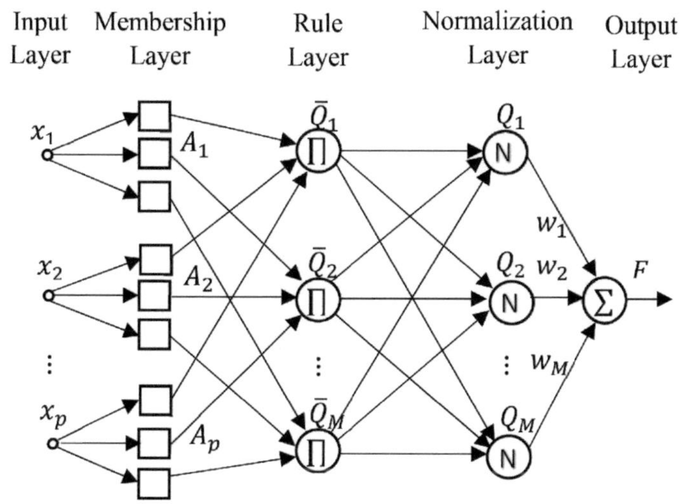

Figure 2.
The structure of a 5-layer ANFIS.

Where σ_{ij} and m_{ij} are the variance and mean of the Gaussian membership function, respectively.

Layer 3 (Rule layer): Fuzzy logic rules are provided as a precondition part in this layer. The input of this layer is the one-dimensional (1D) membership degrees of the associated rule in the second layer. The output of this layer is equivalent to IF in fuzzy rules, which are defined as follows:

$$Q_j^{(3)} = y_j^{(3)} = \prod_i y_{ij}^{(2)}, i = 1, 2, \ldots, p; j = 1, 2, \ldots, M \tag{8}$$

Layer 4 (Normalization layer): The nodes should be normalized in this algorithm. In this layer, the data generated in the previous section are normalized and the output is presented as follows:

$$Q_j^{(4)} = y_j^{(4)} = \frac{y_j^{(3)}}{\left(\sum_{k=1}^{M} y_k^{(3)}\right)}, j = 1, 2, \ldots, M \tag{9}$$

Layer 5 (output layer): The defuzzification is done in this layer. The output of this layer is described as the weighted sum of the previous layers as follows:

$$F = \sum_{k=1}^{M} w_k y_k^{(4)} = W^T Q \tag{10}$$

For estimating the system unknown dynamics, an ANFIS as an intelligent estimator is considered. According to general approximation theory, a neuro-fuzzy estimator $W^{*T}\phi(x)$ is defined as:

$$f(x(t)) = W^{*T}\phi(x(t)) + \delta(x(t)) \tag{11}$$

Where, x and δ denote the system states and approximation error, optimal value W denoted as W^* is obtained by:

$$W^* \triangleq arg\ min \left\{ sup \left| W^T \phi(x) - f(x) \right| \right\} \tag{12}$$

3.2 Assumptions and lemmas

In order to design the desired controller, some assumptions and lemmas are needed, which are introduced in this section.

Assumption 1: The approximation error in (11) and the uncertainties in (4) are bounded and they satisfy these conditions:

$$|\delta| \leq \alpha_1 \tag{13}$$

$$|l(t,x)| = |l_1(t,x)| + |l_2(t,x)| + |l_3(t,x)| \leq \theta_l |x(t)| + |x(t-\tau)| + \alpha_2 \tag{14}$$

Where, α_1, α_2 and θ_l are unknown positive parameters.

Assumption 2: The following conditions are satisfied by unknown functions $x^T(t)x(t)$ and $g^T(x(t))g(x(t))$:

$$x^T(t)x(t) \leq \theta_x |x(t)| \tag{15}$$

$$g^T(x(t))g(x(t)) \leq \theta_g |x(t) + \gamma_1| + \gamma_2 \tag{16}$$

Where γ_1,γ_2 are known positive constants and θ_g and θ_x are unknown positive parameters.

Assumption 3: The time delay is bounded and it satisfies the following conditions:

$$0 \leq \tau \leq \xi, \dot{\tau}(t) \leq \beta < 1 \tag{17}$$

Where, ξ and β are positive constant parameters.

Lemma 1 *(Barbalat's Lemma)* [37]: If conditions i and ii are satisfied, then $\lim_{t \to \infty} g(t) = 0$.

i. $g : R \to R$ is a continuous real function for $t > 0$

ii. $\lim_{t \to \infty} \int_0^t g(\tau) d\tau = L < \infty$.

Lemma 2 [38]: If $\Pi \in R^{n \times n}$ is the positive definite matrix, and $a, b \in R^n$ are any vectors, the following inequality is satisfied:

$$2a^T b \leq a^T \Pi a + b^T \Pi^{-1} b \tag{18}$$

4. Fractional-order sliding mode adaptive intelligent controller (FSAIC) for uncertain fractional-order nonlinear systems with time-varying time delays

In designing FSAIC for the described system of Eq. (4), the controller input is composed of three parts:

$$U(t) = U_e + U_i + U_R \tag{19}$$

Where, U_e is sliding surface feedback and state feedback control:

$$U_e = -kx(t) - \gamma s(t) \tag{20}$$

In order to estimate uncertainty, disturbance and unknown nonlinearities, U_i is approximated by ANFIS approximator, as described and explained in the previous section:

$$U_i = -\widehat{W_f}^T \phi(x(t)) \tag{21}$$

U_R is the robust control signal and represented as:

$$U_R = U_{R1} + U_{R2} \tag{22}$$

Where U_{R1}, U_{R2} can be obtained as follows:

$$U_{R1} = -[\widehat{\alpha}_1 + \widehat{\alpha}_2] - \widehat{\theta}_l|x(t)|\ sgn\ (s(t)) - \left[\widehat{\theta}_g|x(t) + \gamma_1| + \widehat{\theta}_x|x(t)|\right]\ sgn\ (s(t)) \tag{23}$$

$$U_{R2} = -k_R\ sgn\ (s(t)) \tag{24}$$

Where, $(\widehat{W_f})$ stands for approximated variables and the tangential hyperbolic sliding surface of $s(t)$ is defined as follows:

$$s(t) = tanh\left(D^{q-1}x(t) + kD^{-1}x(t)\right) \tag{25}$$

The following adaptive laws update the parameters in the designed controller as:

$$\dot{\widehat{W}} = k_w Sgn(s)\phi(x(t))sech^2\left(D^{q-1}x(t) + kD^{-1}x(t)\right) \tag{26}$$

$$\dot{\widehat{\theta}}_l = l_l|x(t)|sech^2\left(D^{q-1}x(t) + kD^{-1}x(t)\right) \tag{27}$$

$$\dot{\widehat{\theta}}_g = l_g|x(t) + \gamma_1|sech^2\left(D^{q-1}x(t) + kD^{-1}x(t)\right) \tag{28}$$

$$\dot{\widehat{\theta}}_x = l_x|x(t)|sech^2\left(D^{q-1}x(t) + kD^{-1}x(t)\right) \tag{29}$$

$$\dot{\widehat{\alpha}}_i = k_i\ sgn\ (s)sech^2\left(D^{q-1}x(t) + kD^{-1}x(t)\right), (i = 1, 2) \tag{30}$$

Where, k_w, l_l, l_g, l_x and k_i are positive values to be determined.

Theorem 1: The designed controller in (19) guarantees the asymptotic stability of the nonlinear system (4) using the adaptive laws (26–30) and the sliding surface introduced in (25).

Proof: The Lyapunov function candidate is presented as:

$$V = V_1 + V_2 + V_3 \tag{31}$$

Where,

$$V_1 = |s(t)| \tag{32}$$

$$V_2 = sech^2\left(D^{q-1}x(t) + kD^{-1}x(t)\right)\int_{t-\tau}^{t}\left(g^T(x(r))g(x(r)) + x^T(r)x(r)\right)dr \tag{33}$$

$$V_3 = \frac{1}{2}\left(k^{-1}\tilde{W}^T\tilde{W} + l_g^{-1}\tilde{\theta}_g^{T2} + l_l^{-1}\tilde{\theta}_l^{T2} + l_x^{-1}\tilde{\theta}_x^{T2} + \sum_{i=1}^{2} k_i^{-1}\tilde{\lambda}_i^{T2}\right) \tag{34}$$

Where, $\tilde{W} = W - \widehat{W}$, $\tilde{\alpha}_i = \alpha_i - \widehat{\alpha}_i$ $(i = 1, 2)$, $\tilde{\theta}_g = \theta_g - \widehat{\theta}_g$, $\tilde{\theta}_l = \theta_l - \widehat{\theta}_l$ and $\tilde{\theta}_x = \theta_x - \widehat{\theta}_x$. By taking time derivative of Lyapunov function, V_1 can be written as:

$$\dot{V}_1 = \dot{s}(t)\,sgn\,(s(t)) \tag{35}$$

By deriving from the sliding surface (25), the following equation is obtained:

$$\begin{aligned} \dot{s} &= (D^q x(t) + kx(t))sech^2(D^{q-1}x(t) + kD^{-1}x(t)) \\ &= (f(x) + g(x(t-\tau)) + l(t,x) + U(t) + kx(t))sech^2(D^{q-1}x(t) + kD^{-1}x(t)) \end{aligned} \tag{36}$$

By replacing (36) and (22–24) in (35), $\dot{V}_1$ is rewritten as follows:

$$\begin{aligned} \dot{V}_1 &= \dot{s}(t)\,sgn(s(t)) \\ &= sgn(s(t))\left[(f(x) + g(x(t-\tau)) + l(t,x) + U(t) + kx(t))sech^2(D^{q-1}x(t) + kD^{-1}x(t))\right] \\ &\qquad \le [|l(t,x)| - \eta]sech^2(D^{q-1}x(t) + kD^{-1}x(t)) < 0 \end{aligned} \tag{37}$$

Due to the unknown time delay in functions $g(x(t-\tau))$ and $l(t,x)$, they cannot be used in the design of the controller. Therefore, in order to solve the problem caused by the unknown time delay, the Lyapunov-Krasovskii function V_2 is used. The derivative of this function is expressed as follows:

$$\begin{aligned} \dot{V}_2 = &\{g^T(x(t))g(x(t)) + x^T(t)x(t) - (1-\dot{\tau})\left[g^T(x(t-\tau))g(x(t-\tau)) + x^T(t-\tau)x(t-\tau)\right]\} \\ &\times sech^2(D^{q-1}x(t) + kD^{-1}x(t)) - 2\left[tanh\,(D^{q-1}x(t) + kD^{-1}x(t))\right] \\ &\times\left[(D^q x(t) + kx(t))sech^2(D^{q-1}x(t) + kD^{-1}x(t))\right] \\ &\times\left[\int_{t-\tau}^{t}(g^T(x(r))g(x(r)) + x^T(r)x(r))dr\right] \end{aligned} \tag{38}$$

Therefore,

$$\begin{aligned} \dot{V}_2 = &\{g^T(x(t))g(x(t)) + x^T(t)x(t) - (1-\dot{\tau})[g^T(x(t-\tau))g(x(t-\tau)) \\ &+ x^T(t-\tau)x(t-\tau)]\}sech^2(D^{q-1}x(t) + kD^{-1}x(t)) - \frac{2s(t)\dot{s}(t)V_2}{sech^2(D^{q-1}x(t) + kD^{-1}x(t))} \end{aligned} \tag{39}$$

Summing (37) and (39), it would result in Eq. (40).

$$\begin{aligned} &\dot{V}_1 + \dot{V}_2 = \dot{s}(t)\,sgn(s(t)) \\ &-2\dot{s}(t)\,sgn\,(s(t))\frac{V_2|s(t)|}{sech^2(D^{q-1}x(t) + kD^{-1}x(t))} \\ &+\{g^T(x(t))g(x(t)) + x^T(t)x(t) \\ &-(1-\dot{\tau})[g^T(x(t-\tau))g(x(t-\tau)) \\ &+x^T(t-\tau)x(t-\tau)]\}sech^2(D^{q-1}x(t) + kD^{-1}x(t)) \end{aligned} \tag{40}$$

With rearranging,

$$\begin{aligned}\dot{V}_1 + \dot{V}_2 = \dot{s}(t)\, sgn\,(s(t))\left(1 - 2\frac{V_2|s(t)|}{sech^2\left(D^{q-1}x(t) + kD^{-1}x(t)\right)}\right)\\ +\{g^T(x(t))g(x(t)) + x^T(t)x(t)\\ -(1-\dot{\tau})[g^T(x(t-\tau))g(x(t-\tau))\\ +x^T(t-\tau)x(t-\tau)]\}sech^2\left(D^{q-1}x(t) + kD^{-1}x(t)\right)\end{aligned} \tag{41}$$

Where,

$$\frac{V_2|s(t)|}{sech^2\left(D^{q-1}x(t) + kD^{-1}x(t)\right)} \geq 0 \tag{42}$$

According to (42), Eq. (43) can be concluded:

$$1 - 2\frac{V_2|s(t)|}{sech^2\left(D^{q-1}x(t) + kD^{-1}x(t)\right)} \leq 1 \tag{43}$$

Therefore,

$$\begin{aligned}\dot{V}_1 + \dot{V}_2 \leq \dot{s}(t)\, sgn\,(s(t))\\ +\{g^T(x(t))g(x(t)) + x^T(t)x(t)\\ -(1-\dot{\tau})[g^T(x(t-\tau))g(x(t-\tau))\\ +x^T(t-\tau)x(t-\tau)]\}\, sech^2\left(D^{q-1}x(t) + kD^{-1}x(t)\right)\end{aligned} \tag{44}$$

By replacing (36) and *Assumption 3*, we have:

$$\begin{aligned}\dot{V}_1 + \dot{V}_2 \leq (sgn\,(s(t))[f(x) + g(x(t-\tau)) + l(t,x) + U(t) + kx(t)]\\ +\{g^T(x(t))g(x(t)) + x^T(t)x(t)\\ -(1-\beta)[g^T(x(t-\tau))g(x(t-\tau))\\ +x^T(t-\tau)x(t-\tau)]\}) \times sech^2\left(D^{q-1}x(t) + kD^{-1}x(t)\right)\end{aligned} \tag{45}$$

From (14, 45) can be rewritten as:

$$\begin{aligned}\dot{V}_1 + \dot{V}_2 \leq (sgn(s(t))[f(x) + g(x(t-\tau)) + |x(t-\tau)| + |l_{1,3}(t,x)| + U(t) + kx(t)]\\ +\{g^T(x(t))g(x(t)) + x^T(t)x(t)\\ -(1-\beta)[g^T(x(t-\tau))g(x(t-\tau)) + x^T(t-\tau)x(t-\tau)]\})\\ \times sech^2\left(D^{q-1}x(t) + kD^{-1}x(t)\right)\end{aligned} \tag{46}$$

Using *Lemma 2*, let $a = 1, \Pi = 2(1-\beta)$.If $b = g(x(t-\tau))$ one time and $b = x(t-\tau)$, once again, therefore:

$$2g(x(t-\tau)) < \frac{1}{2(1-\beta)} + 2(1-\beta)g^T(x(t-\tau))g(x(t-\tau)) \tag{47}$$

$$2\,x(t-\tau) < \frac{1}{2(1-\beta)} + 2(1-\beta)x^T(t-\tau)x(t-\tau) \tag{48}$$

It yields that:

$$\begin{aligned}\dot{V}_1 + \dot{V}_2 \le & (\,sgn\,(s(t))[f(x) + \frac{1}{4(1-\beta)} + (1-\beta)g^T(x(t-\tau))g(x(t-\tau)) \\ & + \frac{1}{4(1-\beta)} + (1-\beta)x^T(t-\tau)x(t-\tau) + |l_{1,3}(t,x)| + U(t) \\ & + kx(t)] \\ & + \{g^T(x(t))g(x(t)) + x^T(t)x(t) \\ & -(1-\beta)\left[g^T(x(t-\tau))g(x(t-\tau)) + x^T(t-\tau)x(t-\tau)\right]\}) \\ & \times sech^2\left(D^{q-1}x(t) + kD^{-1}x(t)\right)\end{aligned} \tag{49}$$

By simplifying Eq. (49),

$$\begin{aligned}\dot{V}_1 + \dot{V}_2 \le & (\,sgn\,(s(t))\left[f(x) + \frac{1}{2(1-\beta)} + |l_{1,3}(t,x)| + U(t) + kx(t)\right] \\ & + \{g^T(x(t))g(x(t)) + x^T(t)x(t)\})\times sech^2\left(D^{q-1}x(t) + kD^{-1}x(t)\right)\end{aligned} \tag{50}$$

According to (11) and based on *Assumption 2*, (50) could be rewritten as:

$$\begin{aligned}\dot{V}_1 + \dot{V}_2 \le & (\frac{s(t)}{|s(t)|}\left[W^T\phi(x(t)) + \delta + |l_{1,3}(t,x)| + U(t) + kx(t)\right] + \frac{1}{2(1-\beta)} \\ & + \theta_g|x(t) + \gamma_1| + \gamma_2 + \theta_x|x(t)|) \times sech^2\left(D^{q-1}x(t) + kD^{-1}x(t)\right)\end{aligned} \tag{51}$$

By replacing $U(t)$ from (19) to (24), (51) is represented to:

$$\begin{aligned}\dot{V}_1 + \dot{V}_2 \le & (\,sgn\,(s(t))[W^T\phi(x(t)) + \delta + |l_{1,3}(t,x)| - \widehat{W_f}^T\phi(x(t)) - kx(t) - \gamma s(t) \\ & -[\widehat{\alpha}_1 + \widehat{\alpha}_2] - \widehat{\theta}_l|x(t)|\;sgn\,(s(t)) \\ & -\left[\widehat{\theta}_g|x(t) + \gamma_1| + \widehat{\theta}_x|x(t)|\right]\;sgn\,(s(t)) - k_R\;sgn\,(s(t)) + kx(t)] \\ & + \frac{1}{2(1-\beta)} + \theta_g|x(t,) + \gamma_1| + \gamma_2\theta_x|x(t)|) \times sech^2\left(D^{q-1}x(t) + kD^{-1}x(t)\right)\end{aligned} \tag{52}$$

Using $\frac{s(t)}{|s(t)|} = sgn\,(s(t))$ and also sorting (52):

$$\begin{aligned}\dot{V}_1 + \dot{V}_2 \le & (\,sgn\,(s(t))[\left\{W^T\phi(x(t)) - \widehat{W_f}^T\phi(x(t))\right\} + (\delta + |l_{1,3}(t,x)|) - \gamma s(t) \\ & -[\widehat{\alpha}_1 + \widehat{\alpha}_2] - \widehat{\theta}_l|x(t)|\;sgn\,(s(t))] + \left\{\gamma_2 + \frac{1}{2(1-\beta)} - k_R\right\} \\ & + \left\{\left[\theta_g|x(t) + \gamma_1| + \theta_x|x(t)|\right] - \left[\widehat{\theta}_g|x(t) + \gamma_1| + \widehat{\theta}_x|x(t)|\right]\right\}) \\ & \times sech^2\left(D^{q-1}x(t) + kD^{-1}x(t)\right)\end{aligned} \tag{53}$$

By simplifying (53):

$$
\begin{aligned}
\dot{V}_1 + \dot{V}_2 \le & \left(\left\{\tilde{W}^T \phi(x(t))\ sgn\,(s(t)) - [\widehat{\alpha}_1 + \widehat{\alpha}_2]\ sgn\,(s(t)) - \widehat{\theta}_l |x(t)|\right\} - \gamma s(t)\frac{s(t)}{|s(t)|}\right. \\
& + \left(\frac{1}{2(1-\beta)} + \gamma_2 - k_R\right) + \left(\tilde{\theta}_g |x(t) + \gamma_1| + \tilde{\theta}_x |x(t)|\right) \\
& + (\delta + |l_{1,3}(t,x)|)\ sgn\,(s(t))) \times sech^2\left(D^{q-1}x(t) + kD^{-1}x(t)\right)
\end{aligned} \tag{54}
$$

For simplification, assume:

$$
\frac{1}{2(1-\beta)} + \gamma_2 \le k_R \tag{55}
$$

Therefore,

$$
\begin{aligned}
\dot{V}_1 + \dot{V}_2 \le & \left(\left\{\tilde{W}^T \phi(x(t))\ sgn\,(s(t)) - [\widehat{\alpha}_1 + \widehat{\alpha}_2]\ sgn\,(s(t)) - \widehat{\theta}_l |x(t)|\right\}\right. \\
& - \gamma |s(t)| + \left(\tilde{\theta}_g |x(t) + \gamma_1| + \tilde{\theta}_x |x(t)|\right) \\
& + (\delta + |l_{1,3}(t,x)|)\ sgn\,(s(t))) \times sech^2\left(D^{q-1}x(t) + kD^{-1}x(t)\right)
\end{aligned} \tag{56}
$$

Now the time derivative is taken from the Lyapunov function V_3:

$$
\dot{V}_3 = -\left[k_w{}^{-1}\tilde{W}^T\dot{\widehat{W}} + l_g^{-1}\tilde{\theta}_g^T\dot{\widehat{\theta}}_g + l_l^{-1}\tilde{\theta}_l^T\dot{\widehat{\theta}}_l + l_x^{-1}\tilde{\theta}_x^T\dot{\widehat{\theta}}_x + \sum_{i=1}^{2} k_i^{-1}\tilde{\alpha}^T{}_i\dot{\widehat{\alpha}}_i\right] \tag{57}
$$

By summing (56) and (57):

$$
\begin{aligned}
\dot{V}_1 + \dot{V}_2 + \dot{V}_3 \le & \left(\tilde{W}^T\left[\phi(x(t))\, sgn\,(s(t)) sech^2\left(D^{q-1}x(t) + kD^{-1}x(t)\right) - k_w^{-1}\dot{\widehat{W}}_f\right]\right. \\
& + \tilde{\theta}_g\left[|x(t) + \gamma_1| sech^2\left(D^{q-1}x(t) + kD^{-1}x(t)\right) - l_g^{-1}\dot{\widehat{\theta}}_g\right] \\
& + \tilde{\theta}_x\left[|x(t)| sech^2\left(D^{q-1}x(t) + kD^{-1}x(t)\right) - l_x^{-1}\dot{\widehat{\theta}}_x\right] \\
& + \left[-[\widehat{\alpha}_1 + \widehat{\alpha}_2]\, sgn\,(s(t)) sech^2\left(D^{q-1}x(t) + kD^{-1}x(t)\right) - \sum_{i=1}^{2} k_i^{-1}\tilde{\alpha}^T{}_i\dot{\widehat{\alpha}}_i\right] \\
& + \left[-\widehat{\theta}_l |x(t)| sech^2\left(D^{q-1}x(t) + kD^{-1}x(t)\right) - l_l^{-1}\tilde{\theta}_l^T\dot{\widehat{\theta}}_l\right] \\
& \left. + [(\delta + |l_{1,3}(t,x)|)\, sgn\,(s(t)) - \gamma|s(t)|] \times sech^2\left(D^{q-1}x(t) + kD^{-1}x(t)\right)\right)
\end{aligned} \tag{58}
$$

The adaptive laws (26–30) are replaced in (58) as:

$$
\begin{aligned}
\dot{V}_1 + \dot{V}_2 + \dot{V}_3 \le & (-[\widehat{\alpha}_1 + \widehat{\alpha}_2]\, sgn\,(s(t)) - [\tilde{\alpha}_1 + \tilde{\alpha}_2]\, sgn\,(s(t)) - \tilde{\theta}_l |x(t)| - \widehat{\theta}_l |x(t)| \\
& + (\delta + |l_{1,3}(t,x)|)\, sgn\,(s(t)) \\
& - \gamma |s(t)|) sech^2\left(D^{q-1}x(t) + kD^{-1}x(t)\right) \\
& \le [(\delta + |l_{1,3}(t,x)|) - (\alpha_1 + \alpha_2 + \theta_l |x(t)|) - \gamma |s(t)|] \\
& \times sech^2\left(D^{q-1}x(t) + kD^{-1}x(t)\right)
\end{aligned} \tag{59}
$$

According to *Assumption 1* and $\gamma > 0$:

$$\dot{V}(t) = \dot{V}_1 + \dot{V}_2 + \dot{V}_3 \le [(\delta + |l_{1,3}(t,x)|) - (\widehat{\alpha}_1 + \widehat{\alpha}_2) - \gamma|s(t)|]sech^2(D^{q-1}x(t) + kD^{-1}x(t)) \le 0 \tag{60}$$

Therefore, $\dot{V}(t)$ is negative semi-definite. According to this result, the dynamic system (4) with the designed controller (19) is stable in Lyapunov sense. Next, the asymptotic stability of the designed controller will be investigated. Therefore, (60) can be rewritten as:

$$\Xi(t) \equiv [(\widehat{\alpha}_1 + \widehat{\alpha}_2) - (\delta + |l_{1,3}(t,x)|) - \gamma|s(t)|]sech^2(D^{q-1}x(t) + kD^{-1}x(t)) \le -\dot{V} \tag{61}$$

By integrating $\Xi(t)$ in the [0, t] interval:

$$\int_0^t \Xi(r)dr \le V(0) - V(t) \tag{62}$$

$V(t)$ is bounded and decreasing and $V(0)$ is bounded, therefore:

$$\lim_{t\to\infty} \int_0^t \Xi(r)dr \le \infty \tag{63}$$

The above conditions satisfy Barbalat's Lemma's conditions, therefore, according to Barbalat's Lemma:

$$\lim_{t\to\infty} \Xi(t) = 0 \tag{64}$$

It can be said that by using the proposed controller (19), the trajectories of the system states (4) forced onto the sliding surface defined in (25). Therefore, the fractional-order nonlinear system with time delay, uncertainty and disturbance defined in (4) becomes stable using the proposed control law (18). So, the proof is complete.

5. Fractional-order backstepping adaptive intelligent controller (FBAIC) for uncertain fractional-order nonlinear systems with time-varying time delays

In order to design FBAIC, the number of system states is defined as steps of designing controller. Without loss of generality, in this chapter, it is assumed that the number of system states is 3.

5.1 First step

The first state of system (4) is rewritten with the assumption of $x_1 = y_1$:

$$\begin{aligned} D^{q_1}y_1(t) = & f_1(y_1, \varphi_1(y_1), \varphi_2(y_1(t), y_2(t))) \\ & + g_1(y_1(t-\tau), \varphi_1(y_1(t-\tau)), \varphi_2(y_1(t-\tau), y_2(t-\tau))) + l_1(y,t) + u_1(t) \end{aligned} \tag{65}$$

Where, $x_2 = \varphi_1(y_1)$ and $x_3 = \varphi_2(y_1, y_2)$ are considered as virtual control inputs. In order to control the first state of system (4), which is rewritten as Eq. (65), the first control input is considered as introduced in Eq. (19), as:

$$U_1(t) = U_{1e} + U_{1i} + U_{1R} \tag{66}$$

Where, U_{1e} is comprised as the sum of sliding surface feedback and the state feedback terms in the first subsystem, as follows:

$$U_{1e} = -k_1 y_1(t) - \gamma_1 s_1(t) \tag{67}$$

In order to estimate uncertainty, disturbance and unknown nonlinearities, U_{1i} is used as the output of ANFIS estimator, described earlier as:

$$U_{1i} = -\widehat{W_1}^T \phi(y_1(t)) \tag{68}$$

U_{1R} is the robust control law and represented as:

$$U_{1R} = U_{1R1} + U_{1R2} \tag{69}$$

Where U_{1R1}, U_{1R2} can be presented as follows:

$$U_{1R1} = -[\widehat{\alpha}_{11} + \widehat{\alpha}_{12}] - \widehat{\theta}_{1l} |y_1(t)| \ sgn\ (s_1(t)) - \left[\widehat{\theta}_{1g} |y_1(t) + \gamma_1| + \widehat{\theta}_{1y} |y_1(t)|\right] \ sgn\ (s_1(t)) \tag{70}$$

$$U_{1R2} = -k_{1R} \ sgn\ (s_1(t)) \tag{71}$$

Where, the tangential hyperbolic sliding surface $s_1(t)$ is defined as follows:

$$s_1(t) = tanh\ (D^{q_1-1} y_1(t) + k_1 D^{-1} y_1(t)) \tag{72}$$

The following adaptive laws update the parameters in the designed controller:

$$\dot{\widehat{W}}_1 = k_{1w} Sgn(s_1) \phi(y_1(t)) sech^2 (D^{q_1-1} y_1(t) + k_1 D^{-1} y_1(t)) \tag{73}$$

$$\dot{\widehat{\theta}}_{1l} = l_{1l} |y_1(t)| sech^2 (D^{q_1-1} y_1(t) + k_1 D^{-1} y_1(t)) \tag{74}$$

$$\dot{\widehat{\theta}}_{1g} = l_{1g} |y_1(t) + \gamma_1| sech^2 (D^{q_1-1} y_1(t) + k_1 D^{-1} y_1(t)) \tag{75}$$

$$\dot{\widehat{\theta}}_{1y} = l_{1y} |y_1(t)| sech^2 (D^{q_1-1} y_1(t) + k_1 D^{-1} y_1(t)) \tag{76}$$

$$\dot{\widehat{\alpha}}_{1i} = k_{1i}\, sgn\ (s_1) sech^2 (D^{q_1-1} y_1(t) + k_1 D^{-1} y_1(t)), (i = 1, 2) \tag{77}$$

Where, k_{1w}, l_{1l}, l_{1g}, l_{1y} and k_{1i} are positive values.

Theorem 2: The designed controller in (66) guarantees the asymptotic stability of the nonlinear subsystem (65) using the adaptive laws (73–77) and the sliding surface introduced in (72).

Proof: The Lyapunov function candidate for subsystem (65) is presented to be:

$$V_1 = V_{11} + V_{12} + V_{13} \tag{78}$$

Where:

$$V_{11} = |s_1(t)| \tag{79}$$

$$V_{12} = sech^2\left(D^{q_1}y_1(t) + k_1y_1(t)\right)\int_{t-\tau}^{t} \left(g^2(y_1(r)) + y_1^{\,2}(r)\right)dr \tag{80}$$

$$V_{13} = \frac{1}{2}\left(k_{1w}^{-1}\tilde{W}_1^{\,2} + l_{1g}^{-1}\tilde{\theta}_{1g}^{2} + l_{1l}^{-1}\tilde{\theta}_{1l}^{2} + l_{1y}^{-1}\tilde{\theta}_{1y}^{2} + \sum_{i=1}^{2} k_{1i}^{-1}\tilde{\alpha}^2_{1i}\right) \tag{81}$$

Where, $\tilde{W}_1 = W_1 - \widehat{W}_1$, $\tilde{\alpha}_{1i} = \alpha_{1i} - \widehat{\alpha}_{1i}(i = 1, 2)$, $\tilde{\theta}_{1g} = \theta_{1g} - \widehat{\theta}_{1g}$, $\tilde{\theta}_{1l} = \theta_{1l} - \widehat{\theta}_{1l}$ and $\tilde{\theta}_{1y} = \theta_{1y} - \widehat{\theta}_{1y}$.

Using the same procedure of the proof of *Theorem 1*, it can be shown that the derivation of proposed Lyapunov function (78) is negative definite ($\dot{V}_1 < 0$). Therefore, the subsystem (65) using the proposed controller (66) has asymptotic stability.

5.2 Second step

The fractional-order subsystem $y_1 - y_2$ is presented as follows:

$$\begin{aligned} D^{q_1}y_1(t) =& f_1(y_1, \varphi_1(y_1), \varphi_2(y_1, y_2)) \\ &+ g_1(y_1(t-\tau), \varphi_1(y_1(t-\tau), \varphi_2(y_1(t-\tau), y_2(t-\tau))) \\ &- k_1y_1 - \gamma_1 s_1 - \widehat{W}_1^{\,T}\phi(y_1) - [\widehat{\alpha}_{11} + \widehat{\alpha}_{12}] - \widehat{\theta}_l |y_{11}(t)|\ sgn\,(s_1(t)) \\ &- \left[\widehat{\theta}_{1g}|y_1(t) + \gamma_1| + \widehat{\theta}_{1y}|y_1(t)|\right] sgn\,(s_1(t)) - k_{1R}\ sgn\,(s_1(t)) \end{aligned} \tag{82}$$

$$\begin{aligned} D^{q_2}y_2(t) =& f_2(y_1, y_2 + \varphi_1(y_1), \varphi_2(y_1, y_2)) + l_2(y, t) \\ &+ g_2(y_1(t-\tau), y_2(t-\tau) + \varphi_1(y_1(t-\tau), \varphi_2(y_1(t-\tau), y_2(t-\tau))) + U_2(t) \end{aligned} \tag{83}$$

Where, $y_2 = x_2 - \varphi_1(y_1)$ and the virtual control $\varphi_1(y_1)$ is an estimation of x_2.

In order to control the subsystem $y_1 - y_2$ (82) and (83), which is rewritten above, the second control input (19) will be used:

$$U_2(t) = U_{2e} + U_{2i} + U_{2R} \tag{84}$$

Where, U_{2e} includes sliding feedback and state feedback control terms, as:

$$U_{2e} = -k_2y_2(t) - \gamma_2 s_2(t) \tag{85}$$

In order to estimate uncertainty, disturbance and unknown nonlinearities, U_{2i} is used as the output of the ANFIS approximator, as:

$$U_{2i} = -\widehat{W}_2^{\,T}\phi(y_2(t)) \tag{86}$$

U_{2R} is a robust controller and represented as:

$$U_{2R} = U_{2R1} + U_{2R2} \tag{87}$$

Where U_{2R1}, U_{2R2} can be presented as follows:

$$U_{2R1} = -[\widehat{\alpha}_{21} + \widehat{\alpha}_{22}] - \widehat{\theta}_{2l}|y_2(t)|\ sgn\,(s_2(t)) - \left[\widehat{\theta}_{2g}|y_2(t) + \gamma_2| + \widehat{\theta}_{2y}|y_2(t)|\right]\ sgn\,(s_2(t)) \tag{88}$$

$$U_{2R2} = -k_{2R}\ sgn\,(s_2(t)) \tag{89}$$

Where, the tangential hyperbolic sliding surface $s_2(t)$ is defined as follows:

$$s_2(t) = tanh\left(D^{q_2-1}y_2(t) + k_2 D^{-1}y_2(t)\right) \tag{90}$$

The following adaptive laws update the parameters in the designed controller:

$$\dot{\widehat{W}}_2 = k_{2w}Sgn(s_2)\phi(y_2(t))sech^2\left(D^{q_2-1}y_2(t) + k_2 D^{-1}y_2(t)\right) \tag{91}$$

$$\dot{\widehat{\theta}}_{2l} = l_{2l}|y_2(t)|sech^2\left(D^{q_2-1}y_2(t) + k_2 D^{-1}y_2(t)\right) \tag{92}$$

$$\dot{\widehat{\theta}}_{2g} = l_{2g}|y_2(t) + \gamma_2|sech^2\left(D^{q_2-1}y_2(t) + k_2 D^{-1}y_2(t)\right) \tag{93}$$

$$\dot{\widehat{\theta}}_{2y} = l_{2y}|y_2(t)|sech^2\left(D^{q_2-1}y_2(t) + k_2 D^{-1}y_2(t)\right) \tag{94}$$

$$\dot{\widehat{\alpha}}_{2i} = k_{2i}\,sgn\,(s_2)sech^2\left(D^{q_2-1}y_2(t) + k_2 D^{-1}y_2(t)\right), (i = 1, 2) \tag{95}$$

Where, k_{2w}, l_{2l}, l_{2g}, l_{2y} and k_{2i} are positive values.

Theorem 3: The designed controller in (84) guarantees the asymptotic stability of the nonlinear subsystem (82) and (83) using the adaptive laws (91–95) and the sliding surface introduced in (90).

Proof: The Lyapunov function candidate for subsystem (82) and (83) is presented to be:

$$V_2 = V_{21} + V_{22} + V_{23} \tag{96}$$

Where:

$$V_{21} = V_{11} + |s_2(t)| \tag{97}$$

$$V_{22} = V_{12} + sech^2\left(D^{q_2}y_2(t) + k_2 y_2(t)\right)\int_{t-\tau}^{t}\left(g^2(y_2(r)) + {y_2}^2(r)\right)dr \tag{98}$$

$$V_{23} = V_{13} + \frac{1}{2}\left(k_{2w}^{-1}\tilde{W}_2^{\ 2} + l_{2g}^{-1}\tilde{\theta}_{2g}^2 + l_{2l}^{-1}\tilde{\theta}_{2l}^2 + l_{2y}^{-1}\tilde{\theta}_{2y}^2 + \sum_{i=1}^{2} k_{2i}^{-1}\tilde{\alpha}^2{}_{2i}\right) \tag{99}$$

Where, $\tilde{W}_2 = W_2 - \widehat{W}_2$, $\tilde{\alpha}_{2i} = \alpha_{2i} - \widehat{\alpha}_{2i}$, $(i = 1, 2)$, $\tilde{\theta}_{2g} = \theta_{2g} - \widehat{\theta}_{2g}$, $\tilde{\theta}_{2l} = \theta_{2l} - \widehat{\theta}_{2l}$ and $\tilde{\theta}_{2y} = \theta_{2y} - \widehat{\theta}_{2y}$.

Using the same procedure of the proof of *Theorem 1*, it can be shown that the derivation of proposed Lyapunov function (96) is negative definite ($\dot{V}_2 < 0$). Therefore, the subsystem (82) and (83) using the proposed controller (84) has asymptotic stability.

5.3 Third step

The fractional-order subsystem $y_1 - y_2 - y_3$ is presented as follows:

$$\begin{aligned}D^{q_1}y_1(t) =& f_1(y_1,\varphi_1(y_1),\varphi_2(y_1,y_2))\\&+g_1(y_1(t-\tau),\varphi_1(y_1(t-\tau),\varphi_2(y_1(t-\tau),y_2(t-\tau))) - k_1y_1\\&-\gamma_1s_1 - \widehat{W_1}^T\phi(y_1) - [\widehat{\alpha}_{11}+\widehat{\alpha}_{12}] - \widehat{\theta}_{1l}|y_1(t)|\ sgn\ (s_1(t))\\&-\left[\widehat{\theta}_{1g}|y_1(t)+\gamma_1| + \widehat{\theta}_{1y}|y_1(t)|\right] sgn\ (s_1(t)) - k_{1R}\ sgn\ (s_1(t))\end{aligned} \tag{100}$$

$$\begin{aligned}D^{q_2}y_2(t) =& f_2(y_1,y_2+\varphi_1(y_1),\varphi_2(y_1,y_2))\\&+g_2(y_1(t-\tau),y_2(t-\tau)+\varphi_1(y_1(t-\tau),\varphi_2(y_1(t-\tau),y_2(t-\tau)))\\&-\left[\widehat{\theta}_{2g}|y_2(t)+\gamma_2| + \widehat{\theta}_{2y}|y_2(t)|\right] sgn\ (s_2(t)) - k_2y_2 - \gamma_2s_2 - \widehat{W_2}^T\phi(y_1,y_2)\\&-[\widehat{\alpha}_{21}+\widehat{\alpha}_{22}] - \widehat{\theta}_{2l}|y_2(t)|\ sgn\ (s_2(t))\end{aligned} \tag{101}$$

$$\begin{aligned}D^{q_3}y_3(t) =& f_3(y_1,y_2+\varphi_1(y_1),y_3+\varphi_2(y_1,y_2))\\&+g_3(y_1(t-\tau),y_2(t-\tau)+\varphi_1(y_1(t-\tau),y_3(t-\tau)+\varphi_2(y_1(t-\tau),y_2(t-\tau)))\\&+l_3(y,t)+U_3(t)\end{aligned} \tag{102}$$

Where, $y_3 = x_3 - \varphi_2(y_1,y_2)$ and the virtual control $\varphi_2(y_1,y_2)$ is an estimation of x_3.

In order to control the subsystem of $y_1 - y_2$ as expressed in (100–102), the third control input as described in Eq. (19) will be used, as:

$$U_3(t) = U_{3e} + U_{3i} + U_{3R} \tag{103}$$

Where, U_{3e} is composed of sliding feedback and state feedback control terms as described in the first subsystem:

$$U_{3e} = -k_3y_3(t) - \gamma_3s_3(t) \tag{104}$$

In order to estimate uncertainty, disturbance and unknown nonlinearities, U_{3i} is used as the output of an ANFIS estimator, as:

$$U_{3i} = -\widehat{W_3}^T\phi(y_3(t)) \tag{105}$$

U_{2R} is a robust controller and represented as:

$$U_{3R} = U_{3R1} + U_{3R2} \tag{106}$$

Where U_{2R1}, U_{2R2} can be presented as follows:

$$U_{3R1} = -[\widehat{\alpha}_{31}+\widehat{\alpha}_{32}] - \widehat{\theta}_{3l}|y_3(t)|\ sgn\ (s_3(t)) - \left[\widehat{\theta}_{3g}|y_3(t)+\gamma_3| + \widehat{\theta}_{3y}|y_3(t)|\right]\ sgn\ (s_3(t)) \tag{107}$$

$$U_{3R2} = -k_{3R}\ sgn\ (s_3(t)) \tag{108}$$

Where, $s_3(t)$ is defined as follows:

$$s_3(t) = tanh\ (D^{q_3-1}y_3(t) + k_3D^{-1}y_3(t)) \tag{109}$$

The following adaptive laws update the parameters in the designed controller:

$$\dot{\widehat{W}}_3 = k_{2w} Sgn(s_3)\phi(y_3(t))sech^2\left(D^{q_3-1}y_3(t) + k_3 D^{-1}y_3(t)\right) \tag{110}$$

$$\dot{\widehat{\theta}}_{3l} = l_{3l}|y_3(t)|sech^2\left(D^{q_3-1}y_3(t) + k_3 D^{-1}y_3(t)\right) \tag{111}$$

$$\dot{\widehat{\theta}}_{3g} = l_{3g}|y_3(t) + \gamma_3|sech^2\left(D^{q_3-1}y_3(t) + k_3 D^{-1}y_3(t)\right) \tag{112}$$

$$\dot{\widehat{\theta}}_{3y} = l_{3y}|y_3(t)|sech^2\left(D^{q_3-1}y_3(t) + k_3 D^{-1}y_3(t)\right) \tag{113}$$

$$\dot{\widehat{\alpha}}_{3i} = k_{3i}\, sgn\,(s_3)sech^2\left(D^{q_3-1}y_3(t) + k_3 D^{-1}y_3(t)\right), (i = 1, 2) \tag{114}$$

Where, k_{3w}, l_{3l}, l_{3g}, l_{3y} and k_{3i} are positive values.

Theorem 3: The designed controller in (103) guarantees the asymptotic stability of the nonlinear subsystem (100–102) using the adaptive laws (110–115) and the sliding surface introduced in (109).

Proof: The Lyapunov function candidate for subsystem (100–102) is presented to be:

$$V_3 = V_{31} + V_{22} + V_{23} \tag{115}$$

Where:

$$V_{31} = V_{21} + |s_3(t)| \tag{116}$$

$$V_{32} = V_{22} + sech^2\left(D^{q_3}y_3(t) + k_3 y_3(t)\right)\int_{t-\tau}^{t}\left(g^2(y_3(r)) + {y_3}^2(r)\right)dr \tag{117}$$

$$V_{23} = V_{23} + \frac{1}{2}\left(k_{3w}^{-1}\tilde{W}_3{}^2 + l_{3g}^{-1}\tilde{\theta}_{3g}^2 + l_{3l}^{-1}\tilde{\theta}_{3l}^2 + l_{3y}^{-1}\tilde{\theta}_{3y}^2 + \sum_{i=1}^{2} k_{3i}^{-1}\tilde{\alpha}^2{}_{3i}\right) \tag{118}$$

Where, $\tilde{W}_3 = W_3 - \widehat{W}_3$, $\tilde{\alpha}_{3i} = \alpha_{3i} - \widehat{\alpha}_{3i}$, $(i = 1, 2)$, $\tilde{\theta}_{3g} = \theta_{3g} - \widehat{\theta}_{3g}$, $\tilde{\theta}_{3l} = \theta_{3l} - \widehat{\theta}_{3l}$ and $\tilde{\theta}_{3y} = \theta_{3y} - \widehat{\theta}_{3y}$.

Using the same procedure of the proof of *Theorem 1*, it can be shown that the derivation of proposed Lyapunov function (115) is negative definite ($\dot{V}_3 < 0$). Therefore, the subsystem (100–102) using the proposed controller (103) has asymptotic stability.

Remark: The presented control algorithm (FBAIC) in this section can be generalized for higher order system as well.

6. Controller design for fractional-order nonlinear systems with time delay in states and control input

Theorem 4: According to [39], the regular form of the nonlinear system (1) can be rewritten as follows:

$$\begin{cases} D^q z_1 = f(z_1) + h(z_1(t-\tau)) + l(z_1, t) + z_2(t - \tau_u) \\ D^q z_2(t) = \tilde{v}(t) \\ z_1(t) = p(t)\; t \in [-\tau, 0] \\ z_2(t) = q(t)\; t \in [-\tau_u, 0] \end{cases} \tag{119}$$

Where, $\tilde{v}(t)$ is the control input vector. Using this transformation, the input time delay is converted to the state time delay. For this system, the sliding surface can be defined as follows:

$$s(t) = tanh\left(D^{q-1}z_1(t) + D^{q-1}z_2(t) + kD^{-1}z_1(t)\right) \tag{120}$$

The control law from FSAIC approximator is presented as follows:

$$\begin{aligned}\tilde{v}(t) = &-\widehat{W_f}^T\phi(z_1(t)) - kz_1(t) - \gamma s(t) - [\widehat{\alpha}_1 + \widehat{\alpha}_2] - \widehat{\theta}_l|z_1(t)|\ sgn\,(s(t)) \\ &-\left[\widehat{\theta}_g|z_1(t) + \gamma_1| + \widehat{\theta}_z|z_1(t)|\right]\ sgn\,(s(t)) - k\ sgn\,(s(t)) - z_2(t-\tau_u)\end{aligned} \tag{121}$$

The control law in FBAIC is presented as follows:

$$\begin{aligned}\tilde{v}_j(t) = &-\widehat{W}_j\phi(z_{1j}(t)) - k_jz_{1j}(t) - \gamma_js_j(t) - [\widehat{\alpha}_{1j} + \widehat{\alpha}_{2j}] - \widehat{\theta}_{lj}|z_{1j}(t)|\ sgn\,(s_j(t)) \\ &-\left[\widehat{\theta}_{gj}\left|z_{1j}(t) + \gamma_{1j}\right| + \widehat{\theta}_{zj}|z_{1j}(t)|\right]\ sgn\,(s_j(t)) - k_{Rj}\,sgn\,(s_j(t)) - z_{2j}(t-\tau_u)\end{aligned} \tag{122}$$

Where, $j = 1, 2, \ldots$ is the degree of the nonlinear system.

Asymptotical stability of the system (1) with above control laws is guaranteed.

Proof: The asymptotic stability of system (119) can be proved in the same procedure as *Theorem* 1 for FSAIC and *Theorem 2* to *Theorem 4* for FBAIC.

7. Simulation results

In this section, the effectiveness of the designed controller has been examined via MATLAB simulations. For this purpose, the fractional-order chaotic Chen system with time-varying delay in states, unknown dynamics of system and control input, with considering uncertainty and disturbances have been considered. Also, in order to show the performance of the designed controllers, the results of FSAIC, FBAIC and Fractional-order Sliding mode Control (FSC) will be compared.

The fractional-order chaotic Chen system [40] with time-varying delay in states, uncertainty of system and control input is defined as:

$$\begin{cases} D^{q_1}x_1(\mathrm{t}) = a(x_2(\mathrm{t}) - x_1(t-\tau_1)) + \Delta g_1(x(t), x(t-\tau)) + d_1(t) + U_1(t-\tau_{u1}) \\ D^{q_2}x_2(\mathrm{t}) = cx_1(t-\tau_2) - kx_1(t)x_3(t) + \Delta g_2(x(t), x(t-\tau)) + d_2(t) + U_2(t-\tau_{u2}) \\ D^{q_3}x_3(\mathrm{t}) = mx_1^2(t) - bx_3(t-\tau_3) + \Delta g_3(x(t), x(t-\tau)) + d_3(t) + U_3(t-\tau_{u3}) \end{cases} \tag{123}$$

Where, the initial conditions $x_0 = [0.5, 0.2, 0.3]$, $q_i = 0.97$ and $(a, b, c, k, m) = (10, 4, 40, 10, 2.5)$ are assumed and also, the time delay in system states and control inputs are as follows:

$$[\tau_1, \tau_2, \tau_3] = \left[\frac{1+t}{1+2t}, 0.5(1+Sin(t)), (1+Cos(t))\right] \tag{124}$$

$$[\tau_{u1}, \tau_{u2}, \tau_{u3}] = \left[\frac{1+3t}{1+8t}, 0.2(1+Cos(t)), 0.4(1+Sin(2t))\right] \tag{125}$$

Uncertainties and external disturbance of the system are assumed as follows:

$$\Delta g_1(x(t), x(t-\tau)) = Sin(x_2(t)x_3(t))Cos(x_1(t-\tau_1)) \tag{126}$$

$$\Delta g_2(x(t), x(t-\tau)) = 0.1Sin(x_1(t)x_3(t))Cos(x_2(t-\tau_2)) \tag{127}$$

$$\Delta g_3(x(t), x(t-\tau)) = 2Sin(x_1(t)x_2(t))Cos(x_3(t-\tau_3)) \tag{128}$$

$$d(t) = 0.8u(t) \tag{129}$$

Where, $u(t)$ is a step function. The controller parameters are adjusted as follows:

$$l_x = [5, 0.4, 0.1], l_l = [3, 0.3, 0.1], l_g = [2, 0.2, 3], k_w = [50,50,50], k_1 = [0.9,0.1, 0.1],$$
$$k_2 = [0.5, 0.1, 0.3], \gamma = [2, 2, 2], k = [50, 50, 50], k_R = [5, 5, 5]. \tag{130}$$

The controller is applied to the system at the moment $t = 20$s, and at $t = 35$s the states delay changes as follows:

$$[\tau'_1, \tau'_2, \tau'_3] = \left[\frac{5+25t}{1+11t}, 3(1+Sin(t)), \frac{5+5t}{1+3t}\right] \tag{131}$$

In $t = 50$s, uncertainties and disturbances are changed to

$$\Delta g'_1(x(t), x(t-\tau)) = 4Sin(x_2(t)x_3(t))Cos(x_1(t-\tau_1)) \tag{132}$$

$$\Delta g'_2(x(t), x(t-\tau)) = 6Sin(x_1(t)x_3(t))Cos(x_2(t-\tau_2)) \tag{133}$$

$$\Delta g'_3(x(t), x(t-\tau)) = 4Sin(x_1(t)x_2(t))Cos(x_3(t-\tau_3)) \tag{134}$$

$$d'(t) = [0.5Sin(t), 0.5Sin(t), 0.5Sin(t)]. \tag{135}$$

At $t = 70$s, the delay in control inputs is switched to

$$[\tau'_{u1}, \tau'_{u2}, \tau'_{u3}] = \left[0.4(1+Sin(2t)), 0.3(1+Sin(2t)), \frac{1+2t}{1+5t}\right]. \tag{136}$$

Chen's chaotic attractor has been shown in **Figure 3**.

The results of the designed controllers (FBAIC and FSAIC) are compared with those of the fractional-order sliding mode control (FSC) in **Figures 4–10**. The

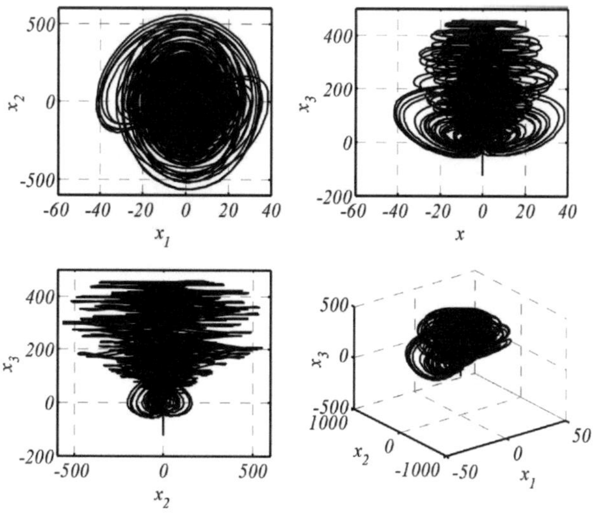

Figure 3.
Fractional order Chen system chaotic attractors.

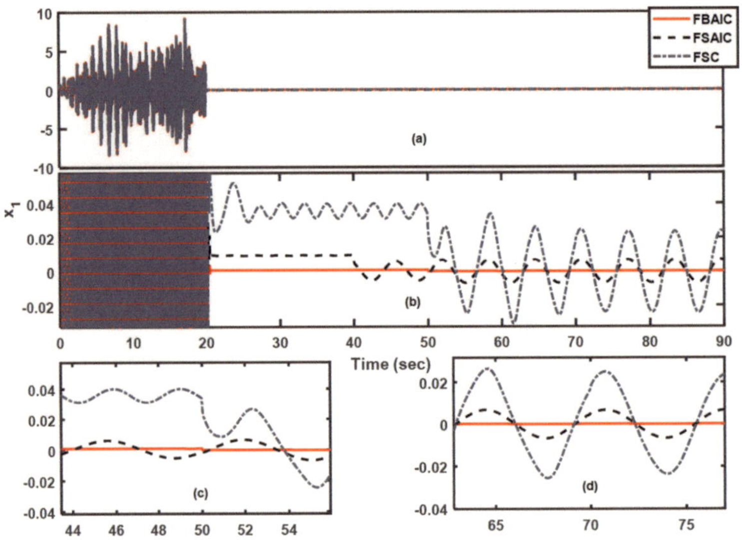

Figure 4.
The first state trajectory of fractional-order Chen system.

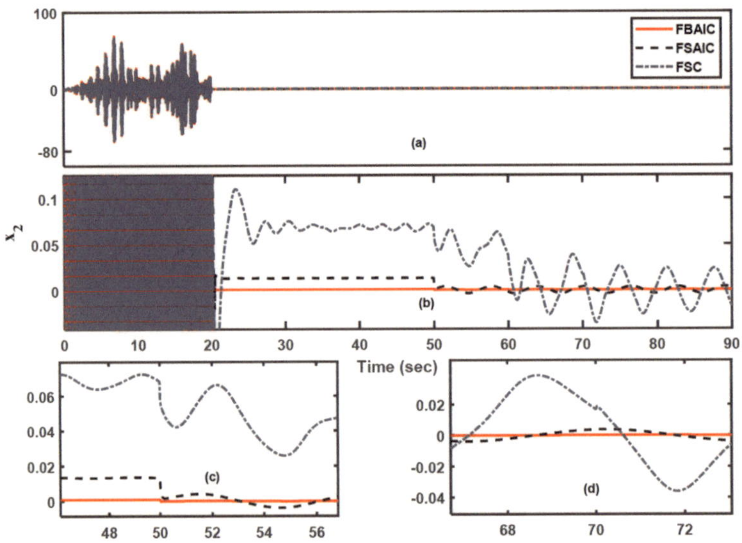

Figure 5.
The second state trajectory of fractional-order Chen system.

system's state trajectories have been shown in **Figures 4–6**. State trajectories have been zoomed out in **Figures 4(b)**, **5(b)** and **6(b)** in the interval [0, 90], while, **Figures 4(c)** to **6(c)** and **4(d)** to **6(d)** focus on the moments of uncertainty and disturbance changes, and variations in time delays, respectively. As seen, the results of the designed controllers have better performance than FSC, as the states do not converge to their desired values and oscillations remain in the state trajectories, even after activating FSC. After applying designed controllers, on the other hand, at $t = 20s$, state trajectories converge to their desired value ($x_d = 0$) in the short time with lower oscillations compared to FSC. However, this is due to employing ANFIS

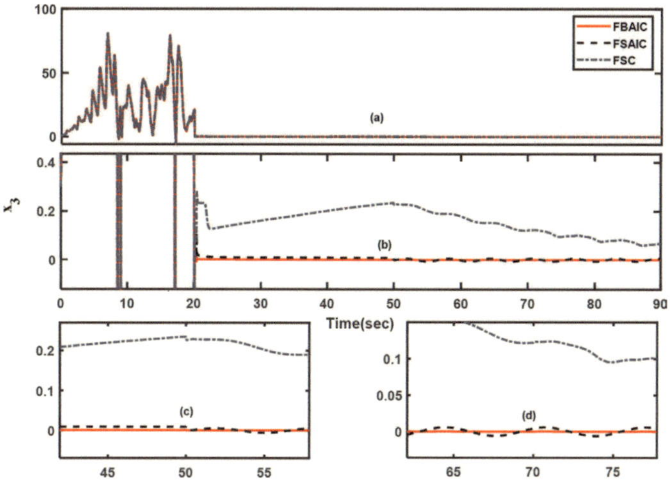

Figure 6.
The third state trajectory of fractional-order Chen system.

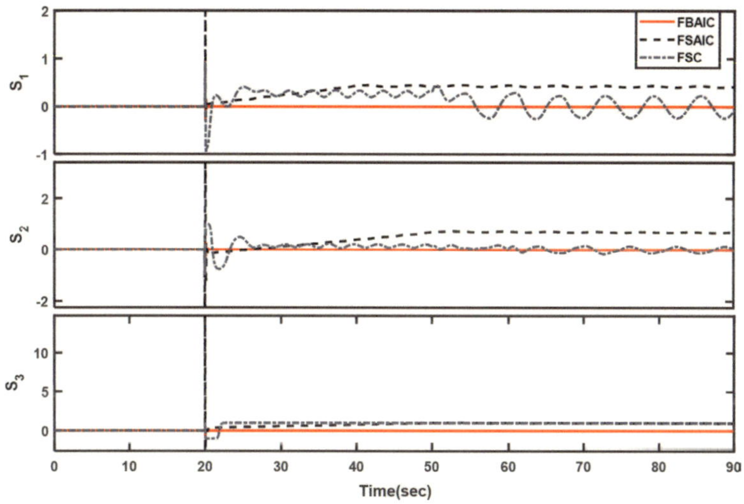

Figure 7.
The sliding surface trajectories.

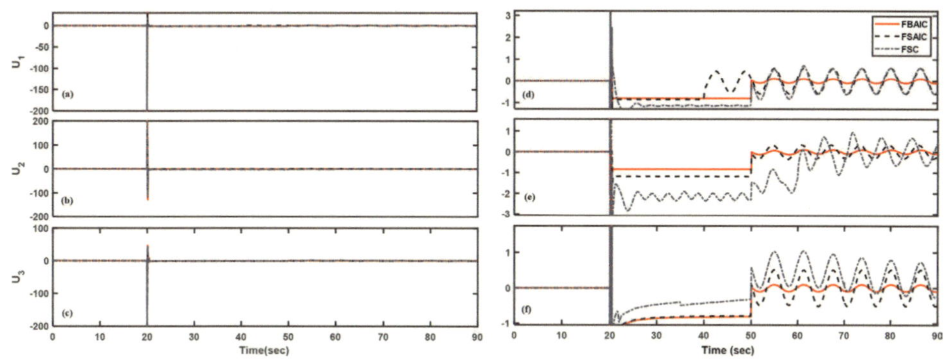

Figure 8.
The input control signals (Right hand: Figures zoomed out for better appearance).

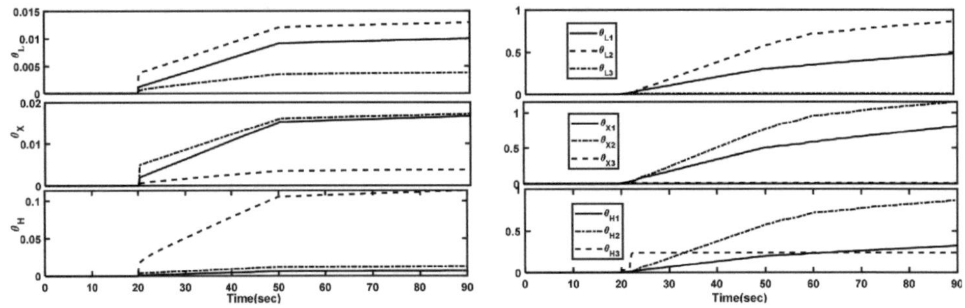

Figure 9.
Left side figure: The updated θ vector parameters of FBAIC; Right side figure: The updated θ vector parameters of FSAIC.

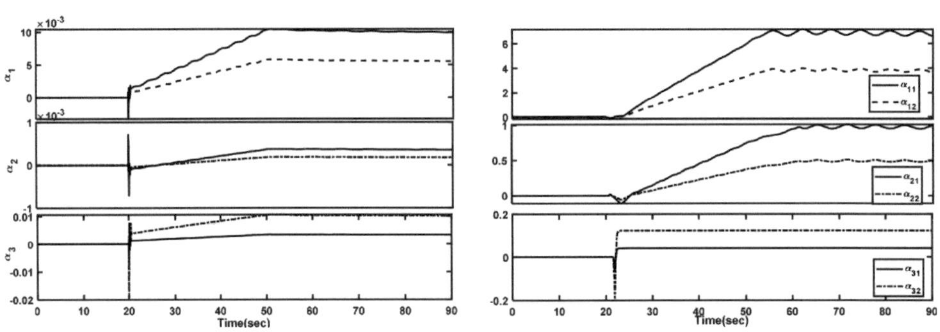

Figure 10.
Left side figure: The updated α vector parameters of FBAIC; Right side figure: The updated α vector parameters of FSAIC.

approximator in the adaptive intelligent robust infrastructure. As another point, one can see in these figures that FBAIC outperform FSAIC, as expected. In addition, in spite of sudden variations in time delays and uncertainty, the proposed controllers have shown good performance and could compensate the variations, rapidly. The same statements presented about state trajectories are true for the sliding surfaces and the control signals, as shown in **Figures** 7 and **8**, respectively. As observed, the control efforts related to the designed controllers have smaller amplitude and lower oscillations compared to the FSC. Also, good capability of the neuro-fuzzy estimator has been reflected in the convergence of the estimated controller parameters, as can be seen in **Figures 9** and **10**. From these figures, it is shown that the neuro-fuzzy estimator has been able to cope with the system properties and has converged, properly, especially in FBAIC.

8. Conclusion

In this chapter, the FBAIC and FSAIC design for fractional-order nonlinear systems with time delays, uncertainty and disturbances was presented. The time delays were separately considered in states, uncertainties and control inputs. Due to the existing time delays, the stability criteria of designed controllers were proved via using Lyapunov-Krasovskii theorem and Barbalat's Lemma. To show the performance of the proposed controllers, FBAIC, FSAIC and FSC were implemented on fractional-order Chen system with uncertainties, disturbances and time-varying delays in its states, uncertainties and control inputs and their results were compared. According to these results, the proposed controllers outperformed FSC in stabilizing the nonlinear systems with uncertainties, disturbances and unknown time-varying delays. As it was clearly seen in the numerical example, despite the sudden variations in time delays, uncertainties and disturbances, regulating the system states has been well done in the shortest time. The obtained control signal also showed less fluctuations and smaller amplitude respecting FSC, while FBAIC overall outperformed FSAIC, as well. Convergence of the proposed controllers' parameters was also representative of the good performance of the neuro-fuzzy estimator to cope with the system properties. Therefore, the hybrid of conventional controller with intelligent controller has improved many problems of conventional controllers, stability, fluctuations of control signals and delayedcompensation.

Author details

Mehdi Dalir and Nooshin Bigdeli*
Control Engineering Department, Imam Khomeini International University, Qazvin, Iran

*Address all correspondence to: n.bigdeli@eng.ikiu.ac.ir

References

[1] Matignon D. Stability properties for generalized fractional differential systems. In: ESAIM: Proceedings. EDP Sciences; 1998

[2] Busłowicz M, Kaczorek T. Recent developments in theory of positive discrete-time linear systems with delays - Stability and robust stability. Measurements, Automatics, Control. 2004;**10**:9-12

[3] Samie H, Moulay E, Coirault P, Vauzelle R. Power control in wireless cellular networks with a time-varying delay. Automatica. 2017;**83**:179-185

[4] Lee J, Chang PH, Jin M. Adaptive integral sliding mode control with time-delay estimation for robot manipulators. IEEE Transactions on Industrial Electronics. 2017;**64**(8):6796-6804

[5] Huynh B, Tjahjowidodo T, Zhong Z-W, Wang Y, Srikanth N. Comparative study of full state variables and time-delay coordinates OGY chaos control for bi-stable vortex induced vibration energy harvesters. In: 2017 IEEE International Conference on Advanced Intelligent Mechatronics (AIM). Munich, Germany: IEEE; 2017

[6] Sun K, Qiu J, Karimi HR, Fu Y. Event-triggered robust fuzzy adaptive finite-time control of nonlinear systems with prescribed performance. IEEE Transactions on Fuzzy Systems. 2020; **29**(6):1460-1471

[7] Sun P, Song X, Song S, Stojanovic V. Composite adaptive finite-time fuzzy control for switched nonlinear systems with preassigned performance. International Journal of Adaptive Control and Signal Processing. 2023; **37**(3):771-789

[8] He Y, Chang X-H, Wang H, Zhao X. Command-filtered adaptive fuzzy control for switched MIMO nonlinear systems with unknown dead zones and full state constraints. International Journal of Fuzzy Systems. 2023;**25**(2):544-560

[9] Wang X, Xu B, Guo Y. Fuzzy logic system-based robust adaptive control of AUV with target tracking. International Journal of Fuzzy Systems. 2023;**25**(1): 338-346

[10] Liang H, Liu G, Zhang H, Huang T. Neural-network-based event-triggered adaptive control of nonaffine nonlinear multiagent systems with dynamic uncertainties. IEEE Transactions on Neural Networks and Learning Systems. 2020;**32**(5):2239-2250

[11] Zhou H, Tong S. Adaptive neural network event-triggered output-feedback containment control for nonlinear MASs with input quantization. IEEE Transactions on Cybernetics. 2023; **53**(11):7406-7416

[12] Tong S, Min X, Li Y. Observer-based adaptive fuzzy tracking control for strict-feedback nonlinear systems with unknown control gain functions. IEEE Transactions on Cybernetics. 2020; **50**(9):3903-3913

[13] Huang Z, Bai W, Li T, Long Y, Chen CP, Liang H, et al. Adaptive reinforcement learning optimal tracking control for strict-feedback nonlinear systems with prescribed performance. Information Sciences. 2023;**621**:407-423

[14] Bahraini MS, Mahmoodabadi MJ, Lohse N. Robust adaptive fuzzy fractional control for nonlinear chaotic systems with uncertainties. Fractal and Fractional. 2023;7(6):484

[15] Liu H, Pan Y, Cao J, Wang H, Zhou Y. Adaptive neural network

backstepping control of fractional-order nonlinear systems with actuator faults. IEEE Transactions on Neural Networks and Learning Systems. 2020;**31**(12): 5166-5177

[16] Zhan Y, Sui S, Tong S. Adaptive fuzzy decentralized dynamic surface control for fractional-order nonlinear large-scale systems. IEEE Transactions on Fuzzy Systems. 2021;**30**(8):3373-3383

[17] Zhan Y, Tong S. Adaptive fuzzy output-feedback decentralized control for fractional-order nonlinear large-scale systems. IEEE Transactions on Cybernetics. 2021;**52**(12):12795-12804

[18] Bi W, Wang T, Yu X. Fuzzy adaptive decentralized control for nonstrict-feedback large-scale switched fractional-order nonlinear systems. IEEE Transactions on Cybernetics. 2021;**52**(9): 8887-8896

[19] Bi W. Neural networks adaptive control for fractional-order non-linear system with unmodelled dynamics and actuator faults. IET Control Theory & Applications. 2023;**17**(3):259-269

[20] Song X, Sun P, Song S, Stojanovic V. Finite-time adaptive neural resilient DSC for fractional-order nonlinear large-scale systems against sensor-actuator faults. Nonlinear Dynamics. 2023;**111**(13): 12181-12196

[21] Zouari F, Ibeas A, Boulkroune A, Cao J. Finite-time adaptive event-triggered output feedback intelligent control for noninteger order nonstrict feedback systems with asymmetric time-varying pseudo-state constraints and nonsmooth input nonlinearities. Communications in Nonlinear Science and Numerical Simulation. 2024;**136**:108036

[22] You X, Dian S, Liu K, Guo B, Xiang G, Zhu Y. Command filter-based adaptive fuzzy finite-time tracking control for uncertain fractional-order nonlinear systems. IEEE Transactions on Fuzzy Systems. 2022;**31**(1):226-240

[23] Ma Z, Sun K. Nonlinear filter-based adaptive output-feedback control for uncertain fractional-order nonlinear systems with unknown external disturbance. Fractal and Fractional. 2023;**7**(9):694

[24] Song S, Park JH, Zhang B, Song X. Adaptive hybrid fuzzy output feedback control for fractional-order nonlinear systems with time-varying delays and input saturation. Applied Mathematics and Computation. 2020;**364**:124662

[25] Dong Y, Song S, Song X, Tejado I. Observer-based adaptive fuzzy quantized control for fractional-order nonlinear time-delay systems with unknown control gains. Mathematics. 2024;**12**(2):314

[26] Montazeri M, Yousefi M, Shojaei K, Shahgholian G. Design of fast variable structure adaptive fuzzy control for nonlinear state-delay systems with uncertainty. IETE Journal of Research. 2022;**68**(6):4577-4589

[27] Wang J, Ni Y, Zhang X, Li Z, Wang C, Su C-Y. Adaptive neural indirect inverse control for a class of fractional-order hysteretic nonlinear time-delay systems and its application. IEEE Transactions on Industrial Electronics. 2024. DOI: 10.1109/TIE.2024.3384589

[28] Li Y, Qu F, Tong S. Observer-based fuzzy adaptive finite-time containment control of nonlinear multiagent systems with input delay. IEEE Transactions on Cybernetics. 2020;**51**(1):126-137

[29] Yang C, Zhang X. Adaptive fuzzy control for fractional-order networked

control systems with input time delay and data loss. Journal of Nonlinear Mathematical Physics. 2024;**31**(1):35

[30] Liu H, Pan Y, Cao J, Zhou Y, Wang H. Positivity and stability analysis for fractional-order delayed systems: A TS fuzzy model approach. IEEE Transactions on Fuzzy Systems. 2020; **29**(4):927-939

[31] Yang J, Wang Y, Wang T, Yang X. Fuzzy-based tracking control for a class of fractional-order systems with time delays. Mathematics. 2022;**10**(11):1884

[32] Xing Y, Wang Y. Finite-time adaptive NN backstepping dynamic surface control for input-delay fractional-order nonlinear systems. IEEE Access. 2023;**11**:5206-5214

[33] Yan Y, Zhang H, Sun J, Wang Y. Sliding mode control based on reinforcement learning for TS fuzzy fractional-order multiagent system with time-varying delays. IEEE Transactions on Neural Networks and Learning Systems. 2024;**35**(8):10368-10379

[34] Zhang X, Wang Z-P, Wu H-N, Zhang X-W, Li H-X, Qiao J-F. Robust non-fragile H_∞ fuzzy control for uncertain nonlinear-delayed hyperbolic PDE systems. International Journal of Fuzzy Systems. 2023;**25**(2): 851-867

[35] Parvizian M, Khandani K, Majd VJ. An H∞ non-fragile observer-based adaptive sliding mode controller design for uncertain fractional-order nonlinear systems with time delay and input nonlinearity. Asian Journal of Control. 2021;**23**(1):423-431

[36] Jang J-S. ANFIS: Adaptive-network-based fuzzy inference system. IEEE Transactions on Systems, Man, and Cybernetics. 1993;**23**(3): 665-685

[37] Slotine J-JE, Li W. Applied Nonlinear Control. NJ: Prentice hall Englewood Cliffs; 1991

[38] Meng J, Wang X. Generalized projective synchronization of a class of delayed neural networks. Modern Physics Letters B. 2008;**22**(03):181-190

[39] Perruquetti W, Borne P, Richard J. A generalized regular form for sliding mode stabilization of MIMO systems. In: Proceedings of the 36th IEEE Conference on Decision and Control. San Diego, CA, USA: IEEE; 1997

[40] Li C, Chen G. Chaos in the fractional order Chen system and its control. Chaos, Solitons & Fractals. 2004;**22**(3): 549-554

Section 2

Applications of Fuzzy Logic

Chapter 3

Cascaded-ANFIS and Its Successful Real-World Applications

Yukinobu Hoshino, Namal Rathnayake, Tuan Linh Dang and Upaka Rathnayake

Abstract

The cascaded adaptive neuro-fuzzy inference system (ANFIS) is widely employed for modeling and controlling nonlinear systems. It combines human experience and knowledge with neural networks' learning capability and fuzzy systems' pattern extraction ability. This integration enables the development of effective models across diverse application domains. This chapter introduces the Cascaded-ANFIS algorithm and its case studies. One example of a case study that uses Cascaded-ANFIS is the modeling of the relationship between rainfall and runoff. This relationship is inherently complex and nonlinear, influenced by watershed topography, soil infiltration characteristics, and rainfall patterns. Accurately capturing this relationship is crucial for flood forecasting and water resources management applications. Rainfall data is the primary input variable when employing Cascaded-ANFIS to model the relationship between rainfall and runoff. This encompasses rainfall data with both temporal and spatial resolutions. Runoff data is collected by observing groundwater levels, river water levels, and geographical features of the watershed.

Keywords: cascaded-ANFIS, gradient boosting, rainfall-runoff prediction, hand gestures prediction, image identification, ensemble model, streamflow prediction

1. Introduction

1.1 Overview of ANFIS

The adaptive neuro-fuzzy inference system (ANFIS) is a hybrid intelligence system that combines neural networks' learning abilities with fuzzy logic's qualitative approach. ANFIS was created by Jang in 1993 [1] and has been widely used in different applications because it can model complex, nonlinear relationships. ANFIS operates by mapping inputs through layers to produce an output. The process can be simplified by combining fuzzy if-then rules with a learning algorithm. This algorithm adjusts system parameters to reduce the difference between predicted and actual outputs. The system's architecture typically involves five layers:

Input Layer: This layer passes the input variables to the next layer.

Fuzzification Layer: In this layer, the input variables are transformed into fuzzy sets using membership functions. If bell-shaped membership functions are used, they can be defined as:

$$\mu_{A_i}(x) = \frac{1}{1 + \left|\frac{x - c_i}{a_i}\right|^{2b_i}} \tag{1}$$

where a_i, b_i, and c_i are the parameters of the bell-shaped function.

Rule Layer: Each node in this layer corresponds to a rule and calculates the firing strength of the rule. The output of this layer is the product of the membership values from the previous layer:

$$O_{2,i} = w_i = \mu_{A_i}(x) \cdot \mu_{B_i}(y) \quad \text{for} \quad i = 1, 2 \tag{2}$$

Normalization Layer: This layer normalizes the firing strengths from the rule layer. The normalized firing strength for each rule is calculated as:

$$O_{3,i} = \overline{w}_i = \frac{w_i}{w_1 + w_2} \tag{3}$$

Defuzzification Layer: In this layer, the output of each rule is computed. This is done by multiplying the normalized firing strength with a linear function of the input variables:

$$O_{4,i} = \overline{w}_i f_i = \overline{w}_i (p_i x + q_i y + r_i) \quad \text{for} \quad i = 1, 2 \tag{4}$$

where p_i, q_i, and r_i are the consequent parameters.

Output Layer: This layer sums the outputs of all the rules to produce the final output of the system:

$$O_5 = \sum_{i=1}^{2} \overline{w}_i f_i = \frac{w_1 f_1 + w_2 f_2}{w_1 + w_2} \tag{5}$$

The learning process in ANFIS involves two main steps: forward pass and backward pass. During the forward pass, the input is propagated forward to calculate the firing strengths and the output. The parameters of the consequent functions are then estimated using the least-squares method. The error is reversed during the backwards pass, and the membership function parameters are adjusted using gradient descent. ANFIS is powerful for modeling and controlling complex systems by combining neural and fuzzy systems. Integrating backpropagation with the least squares method helps the system converge quickly and accurately by improving the parameters effectively.

1.2 Evolution and significance of cascaded-ANFIS

Cascaded-ANFIS is a new approach that improves traditional ANFIS models [2]. ANFIS has evolved into a cascaded structure to overcome challenges like high-dimensional data and computational complexity [3]. This allows it to be used in more complex scenarios.

ANFIS combines the benefits of fuzzy logic and neural networks, providing smoothness and adaptability [4]. However, the traditional ANFIS model struggles with high-dimensional input spaces and computational burdens. The Cascaded-ANFIS solves these issues by utilizing multiple layers. Each layer handles a specific input part and helps improve the final result through an iterative optimization process.

The Cascaded-ANFIS model, with its pair selection and training modules, is designed for efficiency and robustness. The pair selection module identifies the best input pairs, while the training module continuously refines the output for optimal performance. This approach ensures computational efficiency and robustness, making it suitable for real-time applications. Moreover, the algorithm can effectively manage the complexity and improve the system's accuracy by employing a cascaded structure.

The parameters are adjusted in each iteration to reduce the difference between the prediction and the actual results in each ANFIS structure. The example has six inputs as shown in **Figure 1**. Hence, six ANIFS structures were used to obtain the outputs for the first iteration. In this case, the number of parameters that have been tuned is 36. Because each ANFIS structure has p, q, and r design parameters and membership parameters (if bell-shaped: 3 parameters), the first iteration has 36 parameters ($2 \times 3 \times 6$ ANFIS structures) for tuning. Another unique parameter will reappear if the system advances to the next iteration. Hence, increasing the number of iterations increases the complexity of the novel ANFIS algorithm. But, the algorithm is designed to operate only one ANFIS structure at a time. Therefore, the complexity remains consistent in a two-input ANFIS structure. However, the number of tuned parameters increases the accuracy of the optimization solution.

Algorithm 1. Cascaded-ANFIS Algorithm.

1: *Initialize* the system with a two-input ANFIS model.
2: *Input* the dataset with multiple variables.
3: *repeat.*
4: Perform pair selection to identify the best input pairs.
5: Train the ANFIS model with the selected input pairs.
6: Generate the output for the current iteration.
7: Evaluate the performance using a predefined metric (e.g., RMSE).
8: Update the parameters of the ANFIS model.
9: If convergence criteria are met, *exit* loop.
10: *until* convergence criteria are met.
11: *Output* the final optimized model.

By integrating pair selection and iterative training, the Cascaded-ANFIS model continuously refines its output, ensuring high accuracy and efficiency. This method also allows for hybrid optimization techniques, such as particle swarm optimization (PSO) [5–8] and genetic algorithm (GA) [9–11], further enhancing the model's performance.

The evolution of ANFIS into the Cascaded-ANFIS framework represents a significant advancement in computational intelligence. It offers a robust solution to the limitations of traditional ANFIS models, enabling the handling of complex, high-dimensional datasets with improved computational efficiency and accuracy. The benefits of the Cascaded-ANFIS model over traditional ANFIS are clear, making it a superior choice for many applications.

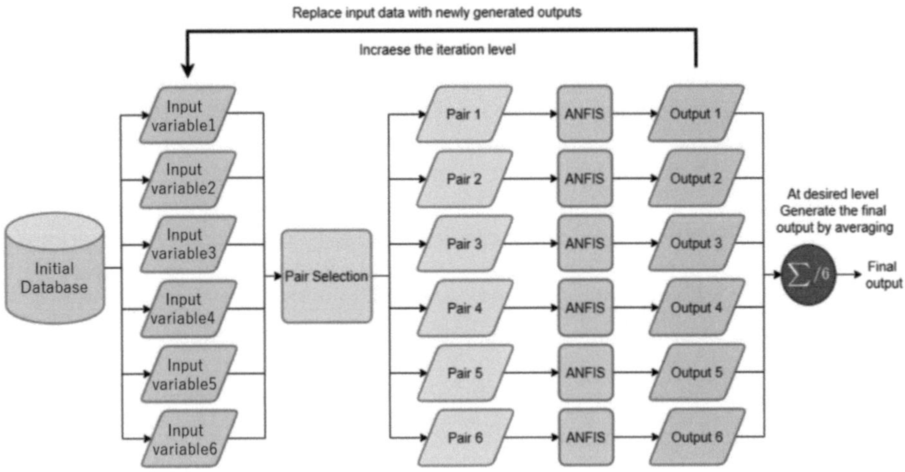

Figure 1.
Example of Cascaded-ANFIS structure.

1.3 Objectives and scope of the chapter

The chapter begins by explaining the traditional ANFIS and showing how it is good at understanding complex, nonlinear relationships [4, 12, 13]. Subsequently, it introduces the concept of the Cascaded-ANFIS, which is designed to improve the effectiveness and flexibility of traditional ANFIS frameworks.

The narrative then explores the development of Cascaded-ANFIS, highlighting the challenges traditional ANFIS methods face, such as the complexity of dimensions and computational requirements [14–16]. Cascaded-ANFIS is important because it helps overcome obstacles with its architecture and optimization process, making it useful in many different areas.

This chapter provides an in-depth analysis of the practical implementation of Cascaded-ANFIS in understanding the complex relationship between rainfall and runoff [17–20]. It underscores the intrinsic complexity inherent in this relationship, accentuating the imperative of precision modeling for flood prognostication and hydrological resource governance tasks [21]. Moreover, it delineates the adept utilization of Cascaded-ANFIS in encapsulating this convoluted nexus, harnessing rainfall data of varying temporal and spatial resolutions.

A systematic explication of the Cascaded-ANFIS algorithm ensues elucidating its procedural intricacies encompassing pair selection, iterative refinement, and performance assessment. Moreover, it expounds upon integrating hybrid optimization methodologies, such as PSO and GA, augmenting the model's efficacy manifold.

Furthermore, the chapter delineates the expansive vista of applications proffered by Cascaded-ANFIS, transcending the realm of rainfall–runoff modeling to encompass domains such as gesture recognition, image classification, and streamflow prediction. It also prognosticates potential avenues for future research and innovation, catalyzing computational intelligence paradigms' evolutionary trajectory.

The chapter furnishes readers with a comprehensive exegesis of Cascaded-ANFIS, spanning its foundational tenets, methodological frameworks, real-world applications, and prospective trajectories. By illuminating its transformative potential, it endeavors to chart a course toward the vanguard of computational intelligence, fostering innovation and advancement across diverse domains.

2. Fundamentals of cascaded-ANFIS

The cascaded adaptive neuro-fuzzy inference system builds on the basic ANFIS model, which combines the learning capabilities of neural networks with the fuzzy logic qualitative approach. Cascaded-ANFIS addresses two main limitations of the traditional ANFIS: the curse of dimensionality and computational complexity [3]. The cascaded approach involves layering multiple ANFIS models to create a more scalable and computationally efficient system.

2.1 Concept and architecture of cascaded-ANFIS

The Cascaded-ANFIS architecture involves a series of ANFIS models arranged in layers. Each layer processes a subset of the input variables, reducing the overall dimensionality handled by each individual ANFIS model. The outputs of one layer serve as inputs to the next layer. This architecture reduces the computational load on each ANFIS model and allows the system to handle more complex and higher-dimensional data more effectively.

2.2 Cascaded approach: structure and advantages

The structure consists of a layered configuration, with each layer in the cascade being an ANFIS model. The first layer processes the initial inputs and passes its outputs to the next layer. A novel pair selection module selects the best input pairs for each ANFIS model, ensuring that each model works with the most relevant variables. Each layer's ANFIS model is also trained independently, simplifying the overall training process.

Cascaded-ANFIS offers several advantages. First, its scalability allows it to handle larger datasets with more input variables by breaking down the problem into smaller parts. Second, it reduces complexity because each individual ANFIS model deals with fewer inputs, making the system less computationally intensive. Finally, its layered approach often results in improved performance and accuracy, as each layer can specialize in processing specific aspects of the input data.

2.3 Learning mechanisms in cascaded-ANFIS

The learning mechanisms in Cascaded-ANFIS involve a combination of backpropagation [22] and least squares estimation [23], similar to traditional ANFIS. However, the cascaded structure introduces additional learning dynamics: Independent training of each ANFIS layer reduces the risk of overfitting and enhances generalization. The cascaded structure facilitates iterative improvement, allowing outputs from one layer to be reprocessed and refined in subsequent layers. Parameters for each ANFIS model are optimized using a combination of gradient descent and least squares methods [24], improving the overall system performance.

By addressing the limitations of traditional ANFIS through a layered and modular approach, Cascaded-ANFIS provides a robust and efficient framework for handling complex, high-dimensional data in various applications.

3. Technical background

The technical background of this study revolves around the ANFIS, first introduced by Jang in 1993. ANFIS combines neural networks (NN) [25] and fuzzy inference systems (FIS) [26] to leverage both strengths. The main advantage of ANFIS is its ability to transform complex systems into simple if-then rules, enabling it to handle nonlinear functions effectively [27]. ANFIS has been widely applied in various research areas, demonstrating powerful results. It combines gradient descent and the least square method to optimize parameter searching, significantly reducing training phase errors.

3.1 Mathematical formulation of cascaded-ANFIS

ANFIS operates through a structured architecture consisting of five layers: input, membership function, fuzzification, normalization, and output layers [28]. For two inputs, x and y, and following the Sugeno FIS model, the ANFIS system applies if-then rules such as

$$f_1 = p_1x + q_1y + r_1 \quad \text{and} \quad f_2 = p_2x + q_2y + r_2, \tag{6}$$

where A_1 and B_1 are fuzzy sets, and p, q, r are design parameters. The first layer, the membership layer, is adaptive and generates membership grades for each input. The second layer computes the firing strengths of the rules. The third layer normalizes these firing strengths, while the fourth layer computes the contribution of each rule to the output. The final layer aggregates these contributions to produce the overall output of the ANFIS model [29].

3.2 Rule generation and parameter optimization

In the rule generation process of ANFIS, the system forms if-then rules based on the input–output data pairs. The membership functions, which define how each input is mapped to a fuzzy set, are determined in the initial phase. Parameter optimization in ANFIS involves adjusting the parameters of these membership functions to minimize the error between the predicted and actual outputs. This optimization typically employs a hybrid learning algorithm that combines the gradient descent method for tuning the parameters of the membership functions and the least squares method for estimating the coefficients of the linear functions in the if-then rules. This dual approach ensures that the ANFIS model accurately captures the underlying patterns in the data.

4. Case study 1: rainfall - runoff prediction

The rainfall–runoff relationship is crucial for understanding hydrological processes and water resource management [30]. Predicting runoff from rainfall data allows for effective planning and mitigation of flood risks [31]. Traditional approaches often struggle with the nonlinear and dynamic nature of this relationship. Advanced machine learning techniques, such as the Cascaded-ANFIS, offer improved accuracy and robustness in modeling these complex interactions.

4.1 Introduction to rainfall–runoff relationship

The rainfall–runoff relationship is a fundamental aspect of hydrology and is pivotal for understanding and managing water resources [32]. This relationship describes how rainfall, as the primary input, translates into runoff, the primary output, within a watershed. Many factors influence it, including soil properties, land use, topography, and climate conditions [33]. Accurate prediction of runoff based on rainfall data is essential for several reasons:

1. Flood Risk Mitigation: Understanding this relationship helps predict potential flood events, allowing for timely and effective flood warning systems and mitigation strategies. Accurate runoff predictions enable authorities to implement necessary measures to protect communities and infrastructure from flood damage [32].

2. Water Resource Management: Effective management of water resources depends on reliable runoff forecasts. This ensures adequate agricultural, industrial, and domestic water supply. Proper planning and allocation of water resources help maintain ecological balance and meet human demands.

3. Hydrological Modeling: The rainfall–runoff relationship forms the basis for hydrological models used in various applications, including watershed management, urban planning, and environmental protection. These models help simulate different scenarios, aiding in decision-making and policy formulation [34].

Traditional approaches to modeling the rainfall–runoff relationship include empirical, conceptual, and physical models [35]. Empirical models rely on historical data and statistical techniques to establish a relationship between rainfall and runoff. Conceptual models represent the catchment as a series of interconnected reservoirs, each simulating different hydrological processes [36]. On the other hand, physical models use detailed representations of physical processes based on the conservation of mass, momentum, and energy [37].

However, these traditional methods often encounter significant challenges:

- Nonlinearity: The relationship between rainfall and runoff is highly nonlinear due to the complex interplay of various hydrological processes. Traditional models may fail to capture these nonlinearities accurately, leading to less reliable predictions [38].

- Temporal and Spatial Variability: Rainfall and runoff exhibit significant variability over time and space. Capturing this variability requires models that can adapt to changing conditions and scale appropriately [39, 40].

- Data Limitations: Traditional models often rely on extensive data for calibration and validation. Often, such data may be incomplete, inaccurate, or unavailable, limiting the model's applicability and accuracy [41, 42].

To address these challenges, advanced machine learning techniques have emerged as powerful tools for modeling the rainfall–runoff relationship. Machine learning algorithms, such as ANN, SVM, and LSTM networks, have demonstrated superior performance in capturing hydrological processes' complex, nonlinear dynamics [43–45].

According to previous research, using ML and AI models for prediction and forecasting still requires further improvements. Therefore, this chapter outlines the methodology and results of utilizing the Cascaded-ANFIS algorithm to address this issue.

4.2 Data collection: rainfall and runoff variables

This study application focused on five major rivers in three countries, as shown in **Figure 2**.

The rivers that were considered in this study can be presented in brief as follows:

1. *Kalu River (Sri Lanka):* The Kalu River in Sri Lanka covers 2766 km^2, with an average annual rainfall of 4000 mm. It flows 129 km from the central highlands to the sea at Kalutara. The Ratnapura area is prone to flooding during the southwest monsoon. Rainfall and water level data from 2000 to 2015 were obtained from the Meteorological Department and the Department of Irrigation.

2. *Kelani River (Sri Lanka):* The Kelani River in Sri Lanka is 145 km long and has a catchment area of 2340 km^2. It provides 80% of Greater Colombo's water supply and receives an average annual rainfall of 3450 mm. Data on daily rainfall and

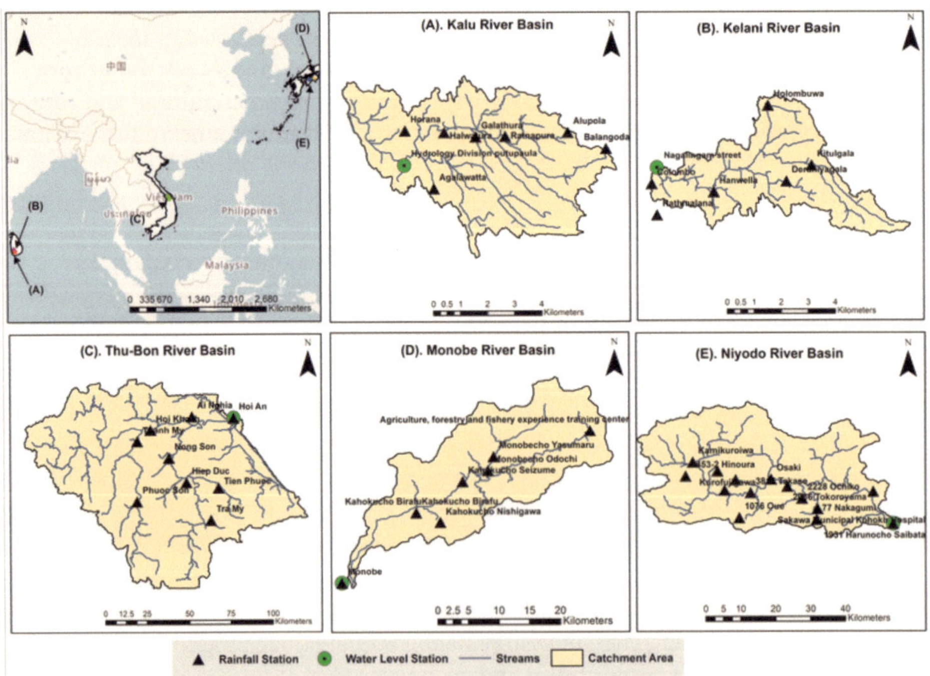

Figure 2.
Rainfall catchment maps of five rivers used in this study: (A) Kalu River, (B) Kelani River, (C) Thu-Bon River, (D) Monobe River, and (E) Niyodo River.

water levels from 2003 to 2017 were obtained from the Meteorological Department and the Department of Irrigation, respectively.

3. *Thu Bon River (Vietnam):* The Thu Bon River in Vietnam stretches 205 km from Ngoc Linh mountain to the South China Sea. It has a catchment area of about 10,000 km^2 and an average discharge of 9100 m^3/s. The river experiences annual flooding from October to December due to Vietnam's rainy season, and data from 2003 to 2007 were obtained from the Vietnam Metrology Institute.

4. *Monobe River (Shikoku Island, Japan):* The Monobe River in Shikoku Island, Japan, originates from Mount Akagiouama in the Tsurugi mountain range. It is approximately 71 km long, with a catchment area of 508.2 km^2 and an average discharge of 24.6 m^3/s, influenced by rainfall. The Japan Meteorological Agency collected data on rainfall and stage measurements from 2010 to 2019.

5. *Niyodo River (Shikoku Island, Japan):* The Niyodo River in Shikoku Island, Japan, originates from Mount Ishizuchi, has a length of approximately 124 km, a basin area of 1560 km^2, and an average discharge of 100 m^3/s. In 1975, the river experienced severe flooding due to a typhoon, prompting close monitoring by Kochi Prefecture. Rainfall and river water level data from 14 rain gauges were obtained from the Japan Meteorological Agency from 2010 to 2019.

4.3 Preprocessing and feature extraction

The preprocessing steps involved cleaning the data to remove any anomalies or missing values, normalizing the variables, and selecting relevant features for the model. Additionally, feature extraction included deriving additional informative metrics from the raw data, such as lagged rainfall variables, to account for the time delay in runoff response. The study considered scenarios with rainfall data from the same day, the previous day, and two days before capturing the temporal dependencies.

4.4 Summary and conclusion of case study 1

This section focuses on five significant rivers across three countries: Kalu and Kelani rivers in Sri Lanka, Thu Bon River in Vietnam, and Monobe and Niyodo rivers in Japan. Detailed data collection was conducted for each river, considering various factors like catchment area, average annual rainfall, and historical data on rainfall and water levels. This comprehensive data collection enabled a robust analysis of the rainfall–runoff relationship for each river, accounting for unique regional characteristics. The manuscript describes the preprocessing steps, which include cleaning the data, normalizing variables, and extracting relevant features such as lagged rainfall variables to capture temporal dependencies.

In conclusion, this section demonstrates the effectiveness of the Cascaded-ANFIS algorithm in predicting runoff from rainfall data. By addressing the limitations of traditional models and leveraging advanced machine learning techniques, the study provides more accurate and reliable predictions, essential for flood risk mitigation and water resource management. The detailed case studies of the five rivers underscore the algorithm's robustness and adaptability to different hydrological contexts. Overall, the study highlights the potential of advanced machine learning techniques in enhancing our understanding and management of complex hydrological processes, paving the way for better-informed decision-making in water resource planning and disaster management [17].

5. Case study 2: hand gestures prediction

5.1 Importance of gesture recognition in human - computer interaction

Gesture recognition is pivotal in human–computer interaction (HCI) as it transforms how users interact with technology by enabling intuitive, natural, and efficient communication with digital systems [46]. By recognizing and interpreting human gestures, particularly hand movements, gesture recognition bridges the gap between human intent and machine response, allowing for a more seamless and immersive user experience. This technology is particularly valuable in enhancing accessibility, as it facilitates non-verbal communication for individuals with disabilities, such as using sign language for the hearing impaired. Additionally, gesture recognition reduces the reliance on traditional input devices, enabling hands-free operation in various contexts, from controlling smart home devices to performing complex tasks in virtual and augmented reality environments [47]. The continuous advancements in machine learning and computer vision drive the development of more accurate and resource-efficient gesture recognition systems, broadening their applications and making them integral to the future of interactive technology [48].

5.2 Dataset description and preprocessing

The dataset used in this study comprises seven distinct hand gesture classes with a total of 350 samples, where each sample is represented by a sequence of 30 frames, summing up to provide a comprehensive view of the gesture [49]. Preprocessing steps are crucial to ensure the quality and reliability of the data. Initially, the dataset undergoes normalization to standardize the range of data values, facilitating better performance in model training. Data augmentation techniques, such as flipping and rotation, are applied to increase the dataset's variability, thus enhancing the model's robustness against different orientations and perspectives of gestures [50]. Additionally, noise reduction methods are employed to minimize any irrelevant variations or artifacts that could obscure the gesture patterns. Finally, the dataset is split into training and testing subsets to enable effective evaluation and validation of the model, ensuring that the model can generalize well to unseen data. These preprocessing steps are essential to prepare the dataset for subsequent machine learning tasks and to improve the accuracy and efficiency of the hand gesture recognition system [51].

5.3 Implementation of ensemble-ANFIS structure for gesture prediction

The results of implementing the Ensemble-ANFIS for gesture prediction showed compelling performance metrics across various evaluation criteria. The overall structure is presented in **Figure 3**. The Ensemble-ANFIS model achieved an average accuracy of 98.9%, surpassing other state-of-the-art algorithms such as XGBoost [52], CatBoost [53], and LightGBM [54], which achieved accuracies of 91.60%, 93.57%, and 87.16%, respectively. This superior accuracy was consistent across different evaluation methods, including 10-fold cross-validation and confusion matrix-based analysis.

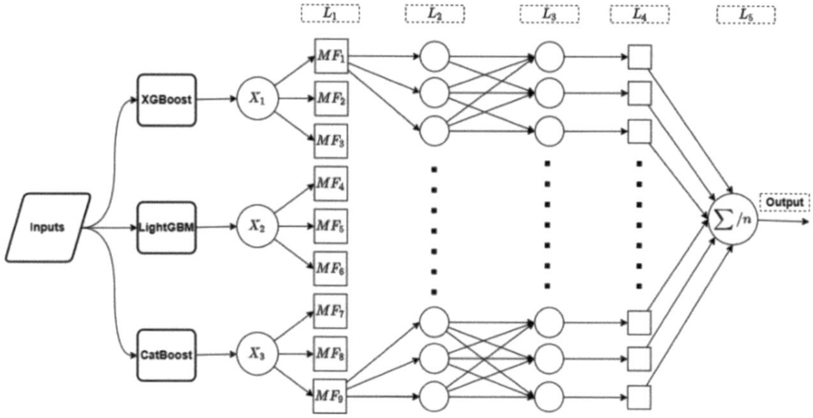

Figure 3.
Ensemble model structure with ANFIS and gradient boosting algorithms.

Specifically, the ANFIS model demonstrated micro-average precision, recall, and F1-score values of 0.99, 0.97, and 0.98, indicating robust performance in correctly identifying and classifying hand gestures. The precision values for individual classes were particularly high, reflecting the model's ability to minimize misclassifications across diverse gesture categories. Overall, these results underscore the effectiveness of Ensemble-ANFIS in leveraging fuzzy inference systems to enhance the accuracy and reliability of gesture prediction in human–computer interaction applications. Implementing Ensemble-ANFIS for gesture prediction presents a promising approach to enhance further the accuracy and robustness of hand gesture recognition systems. Ensemble-ANFIS extends the capabilities of adaptive network-based fuzzy inference systems by integrating multiple ANFIS models in a hierarchical structure. Each level refines the prediction output of the previous level, enabling the system to handle increasingly complex patterns and variations in hand gestures. This approach leverages the iterative learning process of ANFIS to sequentially improve the classification accuracy by refining the membership functions and rules at each stage. By cascading ML models, the system can effectively capture intricate relationships and nuances in gesture data, achieving higher precision and reducing classification errors. **Figure 4** presents the prepossessing, training, and evaluation phase process of the Ensemble-ANFIS model. These processes realize a robust iterative learning process. Moreover, the hierarchical nature of Cascaded-ANFIS allows for incremental learning and adaptation, making it suitable for dynamic environments where gesture patterns may evolve over time. Overall, Ensemble-ANFIS holds great potential in advancing gesture prediction technology, offering a robust framework for real-time applications in human–computer interaction.

5.4 Summary and conclusion of case study 2

This study introduces an innovative approach to hand gesture recognition that integrates ANFIS with gradient boosting methods such as XGBoost, CatBoost, and LightGBM. The ensemble model addresses computational complexity and resource consumption issues associated with traditional CNN-based methods, making it more efficient and effective on low-resource platforms. The proposed method underwent testing on a dataset comprising seven hand gesture classes with 350 samples. Through

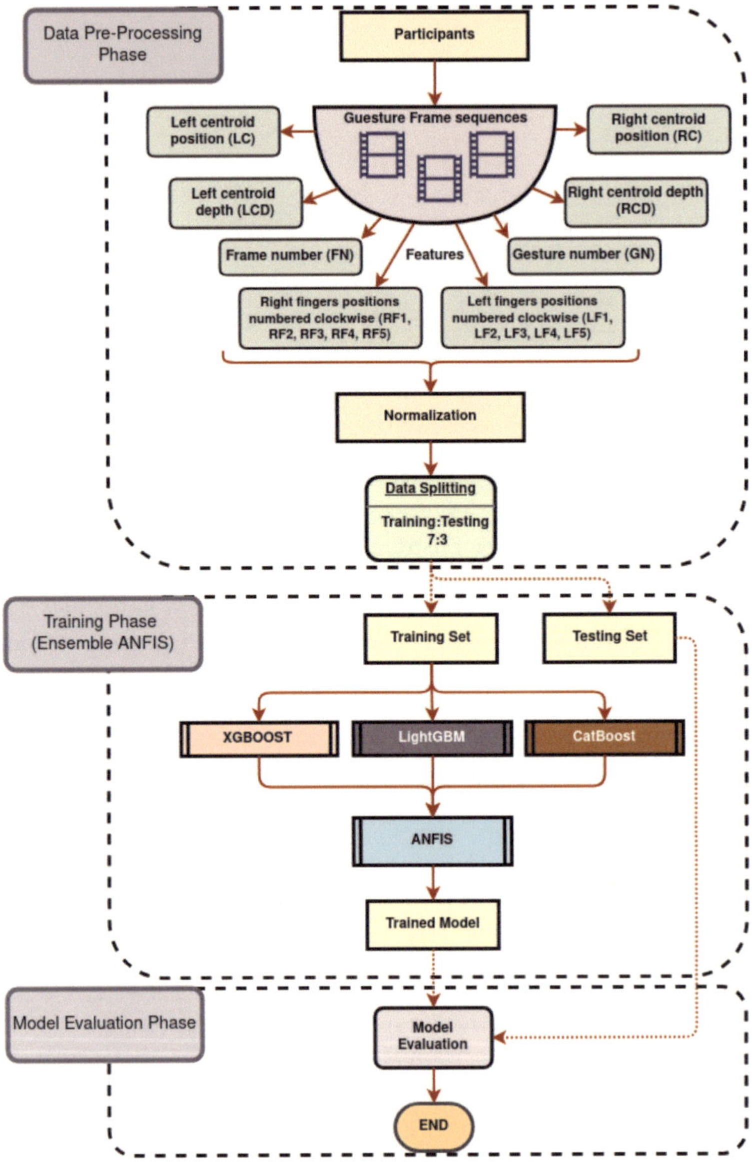

Figure 4.
Prepossessing, training, and evaluation phase process of the Ensemble-ANFIS model.

a rigorous 10-fold cross-validation process, it demonstrated outstanding performance metrics: 0.99 accuracy, 0.99 precision, 0.97 recall, and 0.98 F1 score. Integrating ANFIS into the ensemble model with Cascaded-ANFIS enhances accuracy by effectively handling nonlinear and uncertain variables, surpassing traditional gradient boosting algorithms across various evaluation parameters [55].

6. Case study 3: image identification employed cascaded-ANFIS

Converting features into fuzzy variables allows for constructing flexible fuzzy rules in preprocessing image data, thereby enabling adaptable image recognition [56].

Fuzzy variables effectively manage the ambiguity and uncertainty of data lacking clear boundaries and can flexibly represent complex data characteristics that were difficult to express using traditional rigid numerical methods.

For instance, using vague categories like "slightly reddish green" or "pale blue" instead of strict categories like "red" or "green" for color features makes it easier to capture subtle color differences.

Fuzzy logic can handle complex relationships in data ambiguously, allowing it to flexibly express subtle relationships between features [57, 58]. For example, using fuzzy descriptions such as "rounded corners" or "slightly concave" for shape features enables the flexible identification of minor shape variations. Similarly, by employing fuzzy variables like "slightly rough" or "very smooth" for texture features, texture characteristics can be expressed more naturally.

Additionally, fuzzy rules can account for ambiguous relationships between different features, enabling a flexible rule-based approach that combines multiple features [59]. For instance, the color, shape, and texture features of an image can be integrated to identify complex patterns such as "a smooth object of a slightly bright color on a slightly dark background." This approach allows for the construction of rules that consider the complex features of image data in a holistic manner, significantly improving image recognition accuracy.

Another advantage of fuzzy logic is its robustness to noise, lighting changes, and scale variations [60]. Fuzzy variables can flexibly respond to minor data changes, making them resistant to fluctuations in image data caused by external environmental changes, thus maintaining stable recognition performance [61]. This robustness to noise and environmental changes ensures that the system performs well in image recognition tasks under real-world conditions, providing a sense of security to the audience.

Overall, representing features as fuzzy variables and constructing fuzzy rules enable image recognition systems to manage ambiguity and uncertainty in data effectively and to flexibly identify complex patterns that conventional strict rule-based methods cannot handle. This approach allows the development of flexible image recognition systems capable of comprehensively handling more natural and diverse data.

In preprocessing image data, converting features into fuzzy variables enables the construction of flexible fuzzy rules and facilitates adaptable image classification. This process involves designing membership functions, applying fuzzy rules, and defuzzification.

About Designing Membership Functions: Membership functions are fundamental for converting image features into fuzzy sets [62, 63]. For example, instead of clear categorizations such as "red" and "blue" for color features, we create continuous categories like "slightly reddish" and "pale blue." This allows for the flexible capture of subtle color changes. Similarly, we design membership functions for shape features that can continuously express various shapes, from "perfect circle" to "slightly elliptical." Texture features are also defined from "very rough" to "smooth," and a membership function is designed to capture subtle changes in image texture. By designing membership functions appropriate for each feature, complex information in the image can be managed ambiguously and expressed flexibly.

Fuzzy rules are constructed to express relationships between multiple features using fuzzy variables defined by membership functions [64]. For instance, fuzzy rules might state, "If the image's color is slightly red, the shape is slightly circular, and the texture is slightly smooth, then the image is an apple." This allows for the flexible identification of complex patterns resulting from feature combinations. Fuzzy rules consider subtle relationships between features and possess high discriminatory power even for ambiguous data. Moreover, these rules enable stable discrimination even if features fluctuate slightly due to noise or environmental changes.

Defuzzification is performed to convert fuzzy conclusions obtained by fuzzy rules into concrete decisions [65]. In the defuzzification step, for example, the likelihood of an image being an "apple" is expressed numerically, and the final image classification is based on that numerical value. Common defuzzification methods include the centroid method, which converts the conclusions obtained into concrete class labels and classification results. This makes it possible to make clear decisions even when dealing with ambiguous data.

By designing membership functions, applying fuzzy rules, and defuzzification, it becomes possible to represent features using fuzzy variables flexibly. This approach effectively handles complex and ambiguous image data that cannot be managed by conventional strict rule-based methods and enables flexible image recognition that is resistant to noise and environmental changes. Consequently, it is possible to develop an efficient and highly accurate image recognition system that more naturally reflects the diversity of real-world image data.

6.1 Results analysis and accuracy assessment

We present a series of experiments to develop and test an automated fruit classification system using a novel modified cascaded adaptive network-based fuzzy inference system [66]. The experiments utilized the Fruit-360 dataset, which comprises 131 fruit classes with 90,483 images, **Figure 5**. To classify fruit images effectively, the researchers implemented the Cascaded-ANFIS algorithm, which integrates various image features such as color structure [67], region shape [68], edge histogram [69],

Figure 5.
Examples of 360 fruits image dataset.

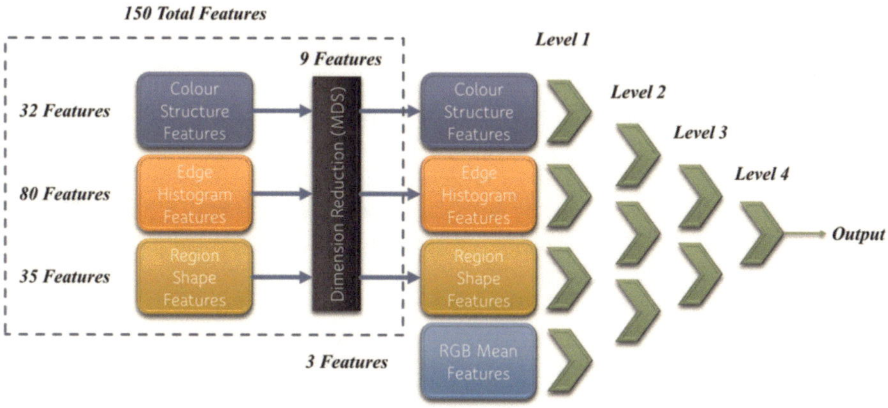

Figure 6.
Cascaded-ANFIS structure for fruits image classification.

and others. The dataset was split into a training set (67,692 images) and a test set (22,688 images), with images being processed to remove the background and standardize the size to 100 × 100 pixels. The classification accuracy of the Cascaded-ANFIS model was evaluated using a confusion matrix, and the performance was compared with existing models like CNN-based algorithms (VGG16, Inception V3, MobileNet, and ResNet18) (**Figure 6**) [70–75].

The results from the experiments showed that the proposed Cascaded-ANFIS algorithm achieved a remarkable classification accuracy of 98.36% on the Fruit-360 dataset. The algorithm also demonstrated high precision, recall, and F-score, with weighted values of 0.9843, 0.9841, and 0.9840, respectively. Compared to other state-of-the-art methods, such as CNN-based models, which ranged from 89.10% to 99.27% accuracy, the Cascaded-ANFIS model proved to be highly effective. The inclusion of multiple image features and the use of feature reduction techniques contributed to the robustness and efficiency of the classification system. The computational efficiency of the Cascaded-ANFIS was also highlighted, showing a significant improvement over other models in terms of processing time and resource requirements.

The accuracy of the Cascaded-ANFIS algorithm was thoroughly validated through multiple methods. The confusion matrix provided a detailed breakdown of the classification performance across all 131 classes, indicating a strong consistency in the model's ability to identify different fruit types correctly. The high precision, recall, and F-score values suggest that the model is both precise in its predictions and reliable across various classes. Furthermore, the comparative analysis with other models underlines the effectiveness of the Cascaded-ANFIS algorithm in handling the complexities of the Fruit-360 dataset. The experiments demonstrated that the model not only achieves high accuracy but also does so with lower computational demands, making it a practical solution for real-world applications in automated fruit classification.

6.2 Comparative analysis with conventional image identification techniques

The proposed Cascaded-ANFIS performance for fruit classification was rigorously compared to conventional image identification techniques. This comparative analysis highlighted the advantages of the Cascaded-ANFIS model over traditional methods in terms of accuracy, computational efficiency, and robustness in handling a large and diverse dataset [66].

Traditional image classification techniques generally rely on manually engineered feature extraction followed by a classification step using algorithms like support vector machines (SVMs) [76], k-nearest neighbors (k-NN) [77], and decision trees. These methods often involve feature extraction techniques such as histogram of oriented gradients (HOG) [77], scale-invariant feature transform (SIFT) [78], and local binary patterns (LBP). These conventional approaches typically require significant domain knowledge for effective feature extraction and are limited by their inability to generalize well across different types of image data.

One critical comparison made in the study was between the Cascaded-ANFIS and traditional methods that utilize manually engineered features. The Cascaded-ANFIS integrates multiple image features, including color structure, region shape, and edge histogram, which are automatically processed by the system to enhance classification accuracy.

On the other hand, traditional methods often require manual tuning of feature extraction parameters, which can be time-consuming and less effective in capturing the complex characteristics of fruit images. The Cascaded-ANFIS, with its adaptive feature processing capabilities, showed a clear advantage in automating this process and improving classification performance.

Regarding classifiers, the Cascaded-ANFIS was compared to commonly used machine learning techniques such as SVMs and k-NN. These traditional classifiers typically struggle with large, multi-class datasets like Fruit-360 due to their limited scalability and the need for extensive computational resources for training and classification. The Cascaded-ANFIS demonstrated superior scalability and efficiency, handling the large dataset more effectively and with lower computational costs. The study noted that traditional classifiers often required longer training times and more computational power, a significant limitation for practical applications.

This section also presented a detailed comparison between the Cascaded-ANFIS and several state-of-the-art convolutional neural networks (CNNs) such as VGG16, Inception V3, MobileNet, and ResNet18. While CNNs are known for their powerful image classification capabilities, they often require substantial computational resources for training and inference. The Cascaded-ANFIS, in contrast, achieved competitive accuracy with significantly lower computational requirements. The study highlighted that the proposed model reached an accuracy of 98.36%, which is comparable to or even surpasses the performance of CNN-based models that generally achieve accuracies ranging from 89.10% to 99.27%. This makes the Cascaded-ANFIS not only a high-performing but also a cost-effective alternative to CNNs.

The comparative analysis underscored the computational efficiency of the Cascaded-ANFIS, which is critical for real-world applications where computational resources may be limited. The traditional methods, especially those involving CNNs, often require powerful hardware and longer processing times, which may not be feasible for all users. The Cascaded-ANFIS, with its reduced computational complexity and resource requirements, offers a more practical solution for large-scale and real-time fruit classification tasks.

Another key advantage of the Cascaded-ANFIS over traditional methods is its robustness and adaptability to a wide range of image variations. The traditional techniques, with their reliance on manually engineered features, often struggle with variations in lighting, scale, and orientation. The Cascaded-ANFIS, through its integration of diverse image features and adaptive processing, demonstrated a higher level of robustness in dealing with these challenges. This makes it a more versatile and reliable tool for automated image classification in diverse environments.

In conclusion, the comparative analysis has shown that the Cascaded-ANFIS model not only outperforms conventional image identification techniques in terms of accuracy and efficiency but also offers greater robustness and adaptability. This makes it a powerful and practical tool for automated fruit classification. The study's findings underscore the potential of the Cascaded-ANFIS model in real-world applications, particularly in agriculture, food processing, and supply chain management.

7. Conclusion and contribution

In conclusion, the study by Rathnayake et al. [66] presents significant advancements in the field of automated fruit classification through the development of the Cascaded-ANFIS. This innovative approach achieves superior classification accuracy and demonstrates remarkable computational efficiency and robustness across a diverse and extensive dataset. The high accuracy rate of 98.36% underscores the model's efficacy in distinguishing between 131 fruit classes, showcasing its potential for practical applications in areas such as agriculture, food processing, and supply chain management. The study's comparative analysis with traditional and contemporary image identification techniques highlights the Cascaded-ANFIS's advantages in handling large datasets, reducing computational complexity, and enhancing classification performance.

The primary contribution of this work lies in integrating multiple image features into a unified adaptive system, which improves classification accuracy without the need for extensive computational resources typically required by CNNs. This approach addresses the limitations of conventional methods that rely heavily on manually engineered features and struggle with scalability and generalization. Furthermore, the study provides a comprehensive evaluation framework, incorporating metrics such as accuracy, precision, recall, and F1-score, which can serve as a benchmark for future research in the field.

Overall, the Cascaded-ANFIS model represents a significant step forward in automated image classification, offering a practical and efficient solution for large-scale applications. Its ability to maintain high performance with lower computational demands makes it an attractive alternative to more resource-intensive models, paving the way for broader implementation in real-world scenarios. The insights and methodologies presented in this study contribute valuable knowledge to the domain, providing a robust foundation for further exploration and innovation in automated classification systems.

Author contribution

Yukinobu Hoshino was responsible for the overall writing and editing of the manuscript. Namal Rathnayake was in charge of the basic study design, data collection, and analysis. Tuan Linh Dang and Upaka Rathnayake assisted with the study design and provided overall supervision and proofreading of the manuscript.

Author details

Yukinobu Hoshino[1*†], Namal Rathnayake[2†], Tuan Linh Dang[3†]
and Upaka Rathnayake[4†]

1 School of Systems Engineering, Kochi University of Technology, Kami, Kochi, Japan

2 Graduate School of Engineering, University of Tokyo, Bunkyo-ku, Tokyo, Japan

3 School of Information and Communications Technology, Hanoi University of Science and Technology, Hai Ba Trung, Hanoi, Vietnam

4 Faculty of Engineering and Design, Department of Civil Engineering and Construction, Atlantic Technological University, Sligo, Ireland

*Address all correspondence to: hoshino.yukinobu@kochi-tech.ac.jp

† These authors contributed equally.

References

[1] Jang J-SR. ANFIS: Adaptive-network-based fuzzy inference system. IEEE Transactions on Systems, Man, and Cybernetics. 1993;**23**(3):665-685

[2] Rathnayake N, Dang TL, Hoshino Y. A novel optimization algorithm: Cascaded adaptive neuro-fuzzy inference system. International Journal of Fuzzy Systems. 2021;**23**:1-17

[3] Salleh M, Talpur N, Hussain K. Adaptive neuro-fuzzy inference system: Overview, strengths, limitations, and solutions. In: Data Mining and Big Data: Second International Conference, DMBD 2017, Fukuoka, Japan, July 27–August 1, 2017, Proceedings 2. Cham, Switzerland: Springer. 2017. pp. 527-535

[4] Kar S, Das S, Ghosh P. Applications of neuro fuzzy systems: A brief review and future outline. Applied Soft Computing. 2014;**15**:243-259

[5] Le T, Huynh T, Lin C. Self-evolving interval type-2 wavelet cerebellar model articulation control design for uncertain nonlinear systems using PSO. International Journal Of Fuzzy Systems. 2019;**21**:2524-2541

[6] Fazel Zarandi M, Dorry F. A hybrid fuzzy PSO algorithm for solving steelmaking-continuous casting scheduling problem. International Journal Of Fuzzy Systems. 2018;**20**: 219-235

[7] Lin C, Le T. PSO-self-organizing interval type-2 fuzzy neural network for antilock braking systems. International Journal Of Fuzzy Systems. 2017;**19**: 1362-1374

[8] Chen C. Optimization of zero-order TSK-type fuzzy system using enhanced particle swarm optimizer with dynamic mutation and special initialization. International Journal Of Fuzzy Systems. 2018;**20**:1685-1698

[9] Liu T, Zhang W, McLean P, Ueland M, Forbes S, Su S. Electronic nose-based odor classification using genetic algorithms and fuzzy support vector machines. International Journal Of Fuzzy Systems. 2018;**20**:1309-1320

[10] Kalia H, Dehuri S, Ghosh A, Cho S. Surrogate-assisted multi-objective genetic algorithms for fuzzy rule-based classification. International Journal Of Fuzzy Systems. 2018;**20**:1938-1955

[11] Zhang C. Classification rule mining algorithm combining intuitionistic fuzzy rough sets and genetic algorithm. International Journal Of Fuzzy Systems. 2020;**22**:1694-1715

[12] Barati-Harooni A, Najafi-Marghmaleki A, Mohammadi A. ANFIS modeling of ionic liquids densities. Journal Of Molecular Liquids. 2016;**224**: 965-975

[13] Tatar A, Barati-Harooni A, Najafi-Marghmaleki A, Norouzi-Farimani B, Mohammadi A. Predictive model based on ANFIS for estimation of thermal conductivity of carbon dioxide. Journal Of Molecular Liquids. 2016;**224**: 1266-1274

[14] Suchetha N, Nikhil A, Hrudya P. Comparing the wrapper feature selection evaluators on twitter sentiment classification. In: 2019 International Conference on Computational Intelligence In Data Science (ICCIDS). Piscataway, NJ, USA: IEEE. 2019. pp. 1-6

[15] Çatalkaya M, Kalıpsız O, Aktaş M, Turgut U. Data feature selection

methods on distributed big data processing platforms. In: 2018 3rd International Conference on Computer Science and Engineering (UBMK). Piscataway, NJ, USA: IEEE. 2018. pp. 133-138

[16] Li L, Wang X, Liu Z, Xie W. A novel intuitionistic fuzzy clustering algorithm based on feature selection for multiple object tracking. International Journal Of Fuzzy Systems. 2019;**21**:1613-1628

[17] Rathnayake N, Rathnayake U, Chathuranika I, Dang T, Hoshino Y. Cascaded-ANFIS to simulate nonlinear rainfall–runoff relationship. Applied Soft Computing. 2023;**147**:110722

[18] Rathnayake N, Rathnayake U, Dang T, Hoshino Y. Water level prediction using soft computing techniques: A case study in the Malwathu Oya, Sri Lanka. PLoS One. 2023;**18**:e0282847

[19] Rathnayake N, Rathnayake U, Chathuranika I, Dang T, Hoshino Y. Projected water levels and identified future floods: A comparative analysis for Mahaweli River, Sri Lanka. IEEE Access. 2023;**11**:8920-8937

[20] Rathnayake N, Rathnayake U, Dang T, Hoshino Y. Streamflow Prediction Using Cascaded-ANFIS Algorithm in Kelani River, Sri Lanka. ISCIIA. Beijing, China: Beijing Institute of Technology. 2022

[21] Rathnayake N, Rathnayake U, Dang T, Hoshino Y. A cascaded adaptive network-based fuzzy inference system for hydropower forecasting. Sensors. 2022;**22**:2905

[22] Rojas R, Rojas R. The backpropagation algorithm. In: Neural Networks: A Systematic Introduction. New York, NY, USA: Springer. 1996. pp. 149-182

[23] Marquardt D. An algorithm for least-squares estimation of nonlinear parameters. Journal Of The Society For Industrial And Applied Mathematics. 1963;**11**:431-441

[24] Orouskhani M, Mansouri M, Orouskhani Y, Teshnehlab M. A hybrid method of modified cat swarm optimization and gradient descent algorithm for training ANFIS. International Journal Of Computational Intelligence And Applications. 2013;**12**:1350007

[25] Abdi H, Valentin D, Edelman B. Neural networks. Thousand Oaks, CA, USA: Sage; 1999

[26] Zadeh L. Fuzzy logic. Computer. 1988;**21**:83-93

[27] Ye L, Ou X. Spatial-temporal analysis of daily air quality index in the Yangtze River Delta region of China during 2014 and 2016. Chinese Geographical Science. 2019;**29**:382-393

[28] Baghban A. Application of the ANFIS strategy to estimate vaporization enthalpies of petroleum fractions and pure hydrocarbons. Petroleum Science And Technology. 2016;**34**:1359-1366

[29] Shamshirband S, Hadipoor M, Baghban A, Mosavi A, Bukor J, Várkonyi-Kóczy A. Developing an ANFIS-PSO model to predict mercury emissions in combustion flue gases. Mathematics. 2019;**7**:965

[30] Sahoo B, Jha R, Singh A, Kumar D. Long short-term memory (LSTM) recurrent neural network for low-flow hydrological time series forecasting. Acta Geophysica. 2019;**67**:1471-1481

[31] Horritt M, Bates P. Evaluation of 1D and 2D numerical models for predicting

river flood inundation. Journal Of Hydrology. 2002;**268**:87-99

[32] Qi W, Ma C, Xu H, Chen Z, Zhao K, Han H. A review on applications of urban flood models in flood mitigation strategies. Natural Hazards. 2021;**108**:31-62

[33] Van Liew M, Arnold J, Bosch D. Problems and potential of autocalibrating a hydrologic model. Transactions Of The ASAE. 2005;**48**: 1025-1040

[34] Gupta H, Kling H, Yilmaz K, Martinez G. Decomposition of the mean squared error and NSE performance criteria: Implications for improving hydrological modelling. Journal Of Hydrology. 2009;**377**:80-91

[35] Klemes V, Others. Empirical and causal models in hydrology. Scientific Basis Of Water Resource Management. 1982:95-104

[36] Jajarmizadeh M, Harun S, Salarpour M. A review on theoretical consideration and types of models in hydrology. Journal Of Environmental Science And Technology. 2012;**5**:249-261

[37] Devia G, Ganasri B, Dwarakish G. A review on hydrological models. Aquatic Procedia. 2015;**4**:1001-1007

[38] Mathias S, McIntyre N, Oughton R. A study of non-linearity in rainfall-runoff response using 120 UK catchments. Journal Of Hydrology. 2016; **540**:423-436

[39] Ogden F, Julien P. Runoff sensitivity to temporal and spatial rainfall variability at runoff plane and small basin scales. Water Resources Research. 1993;**29**:2589-2597

[40] Emmanuel I, Andrieu H, Leblois E, Janey N, Payrastre O. Influence of rainfall spatial variability on rainfall–runoff modelling: Benefit of a simulation approach? Journal Of Hydrology. 2015; **531**:337-348

[41] Coustau M, Ricci S, Borrell-Estupina V, Bouvier C, Thual O. Benefits and limitations of data assimilation for discharge forecasting using an event-based rainfall–runoff model. Natural Hazards And Earth System Sciences. 2013;**13**:583-596

[42] Sitterson J, Knightes C, Parmar R, Wolfe K, Avant B, Muche M. An Overview of Rainfall-Runoff Model Types. Washington, DC, USA: EPA (U.S. Environmental Protection Agency). 2018

[43] Sharghi E, Nourani V, Najafi H, Molajou A. Emotional ANN (EANN) and wavelet-ANN (WANN) approaches for Markovian and seasonal based modeling of rainfall-runoff process. Water Resources Management. 2018;**32**: 3441-3456

[44] Wang W, Xu D, Chau K, Chen S. Improved annual rainfall-runoff forecasting using PSO–SVM model based on EEMD. Journal Of Hydroinformatics. 2013;**15**:1377-1390

[45] Kratzert F, Klotz D, Brenner C, Schulz K, Herrnegger M. Rainfall–runoff modelling using long short-term memory (LSTM) networks. Hydrology And Earth System Sciences. 2018;**22**:6005-6022

[46] Chakraborty D, Garg D, Ghosh A, Chan J. Trigger detection system for american sign language using deep convolutional neural networks. In: Proceedings of the 10th International Conference on Advances In Information Technology. Chichester, UK: John Wiley & Sons. 2018. pp. 1-6

[47] Ameur S, Khalifa A, Bouhlel M. A novel hybrid bidirectional unidirectional

LSTM network for dynamic hand gesture recognition with leap motion. Entertainment Computing. 2020;**35**:100373

[48] Ansar H, Ksibi A, Jalal A, Shorfuzzaman M, Alsufyani A, Alsuhibany S, et al. Dynamic hand gesture recognition for smart lifecare routines via K-Ary tree hashing classifier. Applied Sciences. 2022;**12**:6481

[49] Dang T, Nguyen H, Dao D, Nguyen H, Luong D, Nguyen B, et al. SHAPE: A dataset for hand gesture recognition. Neural Computing And Applications. 2022;**34**:21849-21862

[50] Gupta H, Chudgar H, Mukherjee S, Dutta T, Sharma K. A continuous hand gestures recognition technique for human-machine interaction using accelerometer and gyroscope sensors. IEEE Sensors Journal. 2016;**16**:6425-6432

[51] Kaczmarek P, Mańkowski T, Tomczyński J. putEMG—A surface electromyography hand gesture recognition dataset. Sensors. 2019;**19**:3548

[52] Chen T, Guestrin C. Xgboost: A scalable tree boosting system. In: Proceedings of the 22nd Acm Sigkdd International Conference on Knowledge Discovery and Data Mining. New York, NY, USA: ACM. 2016. pp. 785-794

[53] Prokhorenkova L, Gusev G, Vorobev A, Dorogush A, Gulin A. CatBoost: Unbiased boosting with categorical features. Advances In Neural Information Processing Systems. 2018;**31**

[54] Ke G, Meng Q, Finley T, Wang T, Chen W, Ma W, et al. Lightgbm: A highly efficient gradient boosting decision tree. Advances In Neural Information Processing Systems. 2017;**30**

[55] Rathnayake N, Linh Dang T, Hoshino Y. Designing and implementation of novel ensemble model based on ANFIS and gradient boosting methods for hand gestures classification. In: Proceedings of the 11th International Symposium on Information and Communication Technology. New York, NY, USA: ACM. 2022. pp. 283-289

[56] Höppner F, Klawonn F, Kruse R, Runkler T. Fuzzy Cluster Analysis: Methods for Classification, Data Analysis and Image Recognition. John Wiley & Sons; 1999

[57] Singh R, Bailey W. Fuzzy logic applications to multisensor-multitarget correlation. IEEE Transactions On Aerospace And Electronic Systems. Chichester, UK: John Wiley & Sons. 1997; **33**:752-769

[58] Zadeh L. The role of fuzzy logic in the management of uncertainty in expert systems. Fuzzy Sets and Systems. 1983; **11**:199-227

[59] Gacto M, Alcalá R, Herrera F. Interpretability of linguistic fuzzy rule-based systems: An overview of interpretability measures. Information Sciences. 2011;**181**:4340-4360

[60] Perumal S, Baharum F. Design and simulation of a circadian lighting control system using fuzzy logic controller for LED lighting technology. Journal Of Daylighting. 2022;**9**:64-82

[61] Angelov P, Lughofer E, Zhou X. Evolving fuzzy classifiers using different model architectures. Fuzzy Sets and Systems. 2008;**159**:3160-3182

[62] Klir G, Yuan B. Fuzzy Sets and Fuzzy Logic. Prentice hall New Jersey; Upper Saddle River, NJ, USA: Prentice Hall. 1995

[63] Dubois D, Prade H. Fundamentals of Fuzzy Sets. Springer Science & Business Media. Berlin, Germany: Springer. 2012

[64] Dubois D, Prade H. What are fuzzy rules and how to use them. Fuzzy Sets and Systems. 1996;**84**: 169-185

[65] Bai Y, Wang D. Fundamentals of fuzzy logic control–fuzzy sets, fuzzy rules and defuzzifications. In: Advanced Fuzzy Logic Technologies In Industrial Applications. Berlin, Germany: Springer. 2006. pp. 17-36

[66] Rathnayake N, Rathnayake U, Dang T, Hoshino Y. An efficient automatic fruit-360 image identification and recognition using a novel modified cascaded-ANFIS algorithm. Sensors. 2022;**22**:4401

[67] Weijer J, Others. Color Features and Local Structure in Images. Universiteit van Amsterdam [Host]. Amsterdam, Netherlands: Universiteit van Amsterdam. 2004

[68] Mingqiang Y, Kidiyo K, Joseph R, Others. A survey of shape feature extraction techniques. Pattern Recognition. 2008;**15**:43-90

[69] Park D, Jeon Y, Won C. Efficient use of local edge histogram descriptor. In: Proceedings of the 2000 ACM Workshops on Multimedia. Amsterdam, Netherlands: Universiteit van Amsterdam. 2000. pp. 51-54

[70] Postalcıoğlu S. Performance analysis of different optimizers for deep learning-based image recognition. International Journal Of Pattern Recognition And Artificial Intelligence. 2020;**34**:2051003

[71] Siddiqi R. Effectiveness of transfer learning and fine tuning in automated fruit image classification. In: Proceedings of the 2019 3rd International Conference on Deep Learning Technologies. 2019. pp. 91-100

[72] Huang Z, Cao Y, Wang T. Transfer learning with efficient convolutional neural networks for fruit recognition. In: 2019 IEEE 3rd Information Technology, Networking, Electronic and Automation Control Conference (ITNEC). 2019. pp. 358-362

[73] Ghosh S, Mondal M, Sen S, Chatterjee S, Roy N, Patnaik S. A novel approach to detect and classify fruits using ShuffleNet V2. In: 2020 IEEE Applied Signal Processing Conference (ASPCON). 2020. pp. 163-167

[74] Latif G, Alsalem B, Mubarky W, Mohammad N, Alghazo J. Automatic fruits calories estimation through convolutional neural networks. In: Proceedings of the 2020 6th International Conference on Computer and Technology Applications. 2020. pp. 17-21

[75] Bayliss J, Gualtieri J, Cromp R. Analyzing hyperspectral data with independent component analysis. In: 26th AIPR Workshop: Exploiting New Image Sources and Sensors. Vol. 3240. 1998. pp. 133-143

[76] Zhang Y, Wu L. Classification of fruits using computer vision and a multiclass support vector machine. Sensors. 2012;**12**:12489-12505

[77] Fadhel M, Hatem A, Alkhalisy M, Awad F, Alzubaidi L. Recognition of the unripe strawberry by using color segmentation techniques. International Journal Of Engineering And Technology (UAE). 2018;**7**:3383-3387

[78] Calonder M, Lepetit V, Strecha C, Fua P. Brief: Binary robust independent elementary features. In: Computer Vision–ECCV 2010: 11th European Conference on Computer Vision, Heraklion, Crete, Greece, September 5–11, 2010, Proceedings, Part IV 11. 2010. pp. 778-792

Chapter 4

Toward a Pre- and Post-Performance Decision Support Platform Based on Fuzzy Logic

Manal Tamir, Fatima Ouzayd, Chiheb Reddouane and Hamid Ech-cheikh

Abstract

Problems related to prioritizing a set of data (methods, risks, solutions, etc.), using the opinions of committee members, call into question the need for effective processing based on theoretical methods developed through literature. Indeed, this work is carried out to respond to the needs identified during the analysis of works in the literature that relate to decision support and test effective theoretical approaches in terms of correcting problems linked to imprecision and the inconsistency of opinions. These needs concern both the improvement of calculation performance (results) and the provision of effective technical solutions to respond to certain adaptation errors of the methods used. Therefore, in this work, we will try to develop an IT platform based on theoretical approaches that have shown certain efficiencies in group or individual decision-making. Moreover, this platform is based on multi-criteria decision analysis and is intended for practical and general use. The developed platform uses fuzzy pairwise comparisons and cross-sorting methods.

Keywords: committee members, improvement, performance, cross-sorting methods, fuzzy logic, optimization algorithm

1. Introduction

Prioritizing data (methods, risks, solutions, etc.) based on committee members' opinions raises concerns about the need for effective processing using theoretical methods from the literature. This study aims to address the issues identified in the literature that predict decision support and test theoretical approaches for correcting problems related to imprecision and inconsistent opinions. These issues involve enhancing calculation performance and providing effective technical solutions to adapt to existing methods. We propose developing an IT platform based on theoretical approaches that have demonstrated efficiency in group or individual decision-making. This platform utilizes multi-criteria decision analysis and is designed for practical, general use, incorporating fuzzy pairwise comparisons and cross-sorting methods. The computation methods employed are either validated in the literature or proposed to improve performance and address specific deficiencies. The chosen cross-sorting method is least-squares logarithmic regression due to its ability to manage inconsistencies. Fuzzy logic is applied using triangular membership functions for initial comparisons made by decision-makers, while linguistic values characterize fuzzy possibilities in the comparison matrices. Finally, an indicator measures the consistency of input data, accounting for the acceptable level of inconsistency.

The calculation of normalized weight values can lead to an irrational ordering of fuzzy number elements. Additionally, the method that employs specific modifications to input comparisons appears complex, non-automatic and does not ensure a high consistency indicator. Therefore, this work proposes a classification method to retain only those input comparisons that adhere to the rational order of weight values. Ultimately, an optimization algorithm is designed to maximize the consistency indicator while allowing limited modifications without significantly altering the original data.

The remainder of this chapter is structured as follows: In Section 2, we start with theoretical approaches and calculation methods adopted in the platform. Then, we propose in Section 3, the Consistency Indicator and its calculation formula and optimization algorithm. In Section 4, we give the Architecture and the design of the Decision Support Platform before presenting interfaces in Section 5.

2. Theoretical approaches and calculation methods

2.1 Cross-sorting methods

The essential functionality of our platform consisting of decision-making assistance mainly involves calculating weights. Moreover, the prioritization of these weights will make it possible to construct the decision according to the objectives to be considered.

In this context, we committed to the choice of the cross-sorting method in a literature evaluation process in order to select the one that showed more performance in relation to the treatment of inconsistency. In other words, an analysis of the contributions and limits, relating to the methods detailed in the state of the art section presented in chapter No. 2, is carried out carefully. As already mentioned before, the extension of certain approaches: Geometric Mean on Lines (G.M.L.) approach, approach based on the analysis of eigenvalues (V.P.), to fuzzy numbers proposed by Buckley [1] has been criticized because of the overestimation of imprecision by the concentrated combination of fuzzy numbers. Also, following the proposal of Boender et al. [2] of their own approach which, by comparing its results with those given by Buckely [1] and Van Laarhoven and Pedrycz [3] through the simulation of the same example from Lootsma [4], we can clearly see the improvement in precision in terms of the calculated weights. Therefore, we will opt for this last calculation approach by imposing the reciprocity constraint on the binary comparison matrices, as well as the independent distribution of the values generated subsequently. A set of constraints are posed and tested by Escobar and Moreno-Jiménez [5].

This section will present the calculation equations and define the variables used. Before detailing the calculations, we will also translate the main data employed.

rijk: Value of comparison ij.

αi: Weight of element i.

βij: Weight of element i in relation to criterion j.

CI: Consistency indicator.

2.2 Fuzzy logic: Choice of membership functions

Fuzzy logic utilizes fuzzy numbers characterized by the distribution of possibilities. The membership function μ defines this distribution by linking numerical elements to degrees of possibility ranging from 0 to 1. Variants of membership functions include triangular, trapezoidal, and Gaussian forms. Fuzzy logic plays a significant role in the literature. In this work, we adopt the triangular function used by FDA et ICH [6] in their extension of the least-squares logarithmic regression to account for inaccuracies. The triangular function is defined by the lower (cl), modal (cm), and upper (cu) values, as illustrated in **Figure 1**.

2.3 Reciprocal matrix

Our calculation approach uses reciprocal matrices (see **Figure 2**) which are characterized by values whose symmetries are equal to their inverses, namely: $\mathbf{r_{ijk}} = \frac{1}{\mathbf{r_{jik}}}$

The adoption of fuzzy logic in the weight calculation part involves a representation of each value in three sub-values as follows:

$$\mathbf{r_{ijk}} = \left(\mathbf{r_{ijk\mathit{l}}}, \mathbf{r_{ijk\mathit{m}}}, \mathbf{r_{ijk\mathit{u}}}\right) \ \mathbf{et}\ \mathbf{r_{jik}} = \frac{1}{\mathbf{r_{ijk}}} = \left(\frac{1}{\mathbf{r_{ijku}}}, \frac{1}{\mathbf{r_{ijkm}}}, \frac{1}{\mathbf{r_{ijkl}}}\right).$$

2.4 Linguistics values

Fuzzy logic's distinct feature is its precision, characterized by partial truth values ranging from 0 to 1, unlike traditional Boolean values of 0 or 1. As noted by Zadeh [7], this framework aligns more closely with human reasoning than binary logic. Zadeh also explored linguistic variables and their role in approximate reasoning, suggesting that treating reality as a linguistic variable concretizes fuzzy logic, where values are expressed as words or phrases.

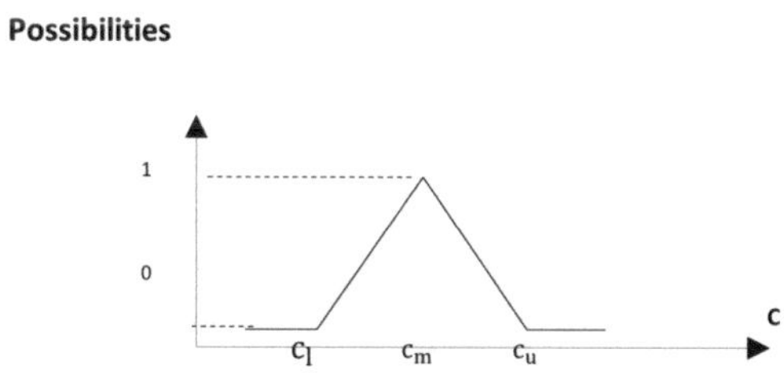

Figure 1.
Triangular function.

$$\begin{pmatrix} 1 & r_{12k} & r_{13k} & \cdots & r_{1nk} \\ 1/r_{12k} & 1 & r_{23k} & \cdots & r_{2nk} \\ 1/r_{13k} & . & 1 & \cdots & r_{3nk} \\ . & . & . & 1 & \cdots \\ . & . & . & . & 1 \end{pmatrix}$$

Figure 2.
Comparison matrix.

Linguistic value	Designation	Its correspondence in fuzzy number
VH	Very high	(7,9,10)
H	High	(6,7,9)
E	Equal	(3,5,7)
L	Low	(1,3,4)
VL	Very low	(0,1,3)

Table 1.
Linguistic values and their translations into fuzzy numbers.

In developing the platform, we will use a linguistic variable to indicate the degree of importance, with five values: very high, high, equal, low, and very low (see **Table 1**). This approach simplifies the evaluation process for users unfamiliar with fuzzy logic by avoiding numerical figures.

2.5 Weight calculation algorithm

Given that the method to be adopted in this work is that proposed by Boender et al. [2] by imposing certain constraints on the matrices of comparison, the weight calculation algorithm is based on the minimization of the least-squares logarithmic regression function given below by formula (Eq. (1)) as well as its conversion using fuzzy logic (Eq. (2)):

$$\sum_{i=1}^{m}\sum_{j=i+1}^{m}\sum_{k\epsilon D_{ij}}\left\{ln\left(r_{ijk}\right)-ln\left(\alpha_I/\alpha_J\right)\right\}^2 \tag{1}$$

$$\sum_{i=1}^{m}\sum_{j=i+1}^{m}\sum_{k\epsilon D_{ij}}\left\{ln\left(r_{jjkl}\right)-ln\left(\alpha_{il}\right)+ln\left(\alpha_{ju}\right)\right\}^2+\left\{ln\left(r_{jjkm}\right)-ln\left(\alpha_{im}\right)+ln\left(\alpha_{jm}\right)\right\}^2$$
$$+\left\{ln\left(r_{jjku}\right)-ln\left(\alpha_{iu}\right)+ln\left(\alpha_{jl}\right)\right\}^2, \tag{2}$$

Then, the corresponding normal equation is fixed by the following formula Eq. (3):

$$\ln(\alpha_i)\sum_{j\neq i}^{m}\delta_{ij}-\sum_{j\neq i}^{m}\delta_{ij}\ln(\alpha_j)=\sum_{j\neq i}^{m}\sum_{k\in D_{ij}}^{m}\ln(r_{ijk})\ i=1,\ldots,m \tag{3}$$

By using the triangular fuzzy numbers, the previous calculation formula will thus be adapted for the calculation of the three fuzzy values of the desired weights. Indeed, three formulas are generated (4), (5) and (6) with three unknown parameters: α_{il}, α_{im}, and α_{iu}. The system of equations is characterized by a dependence between formulas (4) and (6) based on the following initial conditions: $r_{jikl}=\frac{1}{r_{ijku}}$ et $r_{jiku}=\frac{1}{r_{ijkl}}$.

$$\ln(\alpha_{il})\sum_{j\neq i}^{m}\delta_{ij}-\sum_{j\neq i}^{m}\delta_{ij}\ln(\alpha_{ju})=\sum_{j\neq i}^{m}\sum_{k\in D_{ij}}^{m}\ln(r_{ijkl})\ i=1,\ldots,m \tag{4}$$

$$\ln(\alpha_{im})\sum_{j\neq i}^{m}\delta_{ij}-\sum_{j\neq i}^{m}\delta_{ij}\ln(\alpha_{jm})=\sum_{j\neq i}^{m}\sum_{k\in D_{ij}}^{m}\ln(r_{ijkm})\ i=1,\ldots,m \tag{5}$$

$$\ln(\alpha_{iu})\sum_{j\neq i}^{m}\delta_{ij}-\sum_{j\neq i}^{m}\delta_{ij}\ln(\alpha_{jl})=\sum_{j\neq i}^{m}\sum_{k\in D_{ij}}^{m}\ln(r_{ijku})\ i=1,\ldots,m \tag{6}$$

Furthermore, the correction of the normalization proposed by Boender et al. [2] is translated as follows:

$$\tilde{\alpha}_i=(a\ \exp(x_{il}),b\ \exp(x_{im}),a\ \exp(x_{iu}))\ i=1,\ldots,m,\text{Avec}: x_{il}=\ln(\alpha_{il})x_{im}=\ln(\alpha_{im})x_{iu}=\ln(\alpha_{iu}) \tag{7}$$

$$a=\frac{1}{\left(\sum_{i=1}^{m}\exp(x_{il})\sum_{i=1}^{m}\exp(x_{iu})\right)^{\frac{1}{2}}},b=\frac{1}{\sum_{i=1}^{m}\exp(x_{im})} \tag{8}$$

Finally, the weights are calculated as follows:

$$\tilde{\alpha}_i=\begin{pmatrix}\alpha_{il}\\ \alpha_{im}\\ \alpha_{iu}\end{pmatrix}=\begin{bmatrix}\dfrac{exp\,(x_{il})}{\left(\sum_{j=1}^{m}exp\,(x_{jl})\sum_{j=1}^{m}exp\,(x_{ju})\right)^{\frac{1}{2}}}\\ \dfrac{exp\,(x_{im})}{\sum_{j=1}^{m}exp\,(x_{jm})}\\ \dfrac{exp\,(x_{iu})}{\left(\sum_{j=1}^{m}exp\,(x_{jl})\sum_{j=1}^{m}exp\,(x_{ju})\right)^{\frac{1}{2}}}\end{bmatrix}\ i=1,\ldots,m \tag{9}$$

On the other hand, the calculation of weights within the framework of multicriteria analysis results in two levels of calculation. First, the weights of the criteria $\tilde{\alpha}_i=(\alpha_{il},\alpha_{im},\alpha_{iu})$, i = 1, … ., m are calculated from a number of comparison matrices $A_1,\ldots,A_K$ which are equal to the number of evaluators K. Then, the second level of calculation consists of calculating the weights of n elements $\beta_{ij}=\left(\beta_{ijl},\beta_{ijm},\beta_{iju}\right)$ $(i=1,\ldots,n; j=1,\ldots,m)$ with respect to each criterion separately.

Finally, the final weights ω_j j = 1, … ,n are calculated from the aggregation of the previous weights according to the formula below Eq. (10):

$$\omega_j=\sum_{i=1}^{m}\alpha_i\beta_{ij}\ j=1,\ldots,n \tag{10}$$

2.6 Rationality of the order of results

As mentioned in the state-of-the-art section on the use of fuzzy logic in multicriteria decision-making support, the correction of the normalization procedure proposed by Boender et al. [2] does not resolve the issue of the order of the fuzzy elements that make up the results. Moreover, studies addressing this anomaly suggest establishing a set of conditions on the initial comparisons for 3×3 matrices. Consequently, this correction approach results in a heavy computation load for matrices of larger dimensions. Furthermore, the conditions imposed on the evaluators' comparisons may influence their opinions and compromise the objectivity of the decision.

In our decision-making solution, we use a large number of iterations with small perturbations (between −5% and + 5%) applied to the completed comparisons to avoid this ordering issue. This adopted method allows for the selection of both results where the order is respected and those with a significant degree of consistency. A two-value table, 0 and 1, is thus created to classify the iterations. The iteration number is assigned to column 0 if the order is not respected and to column 1 if it is respected. Perturbations are first applied to the entries related to the criteria. If the order is respected for the iteration, the weights are calculated for the candidates, and the iteration number is recorded in column 1 if both orders are respected. Otherwise, it is recorded in column 0. Finally, the consistency indicator is calculated for each iteration that respects the order (see **Figure 3**), and the maximum value of this indicator will be displayed and used as the initial data for our optimization program.

3. Consistency indicator

3.1 Calculation formulas

To evaluate the consistency of the input data, two indicators will be calculated at two levels: the data consistency indicator for each member separately (evaluator's consistency indicator) and the data consistency indicator for the group (group consistency indicator). Furthermore, high values of these indicators ensure that the obtained results come from a consistent dataset. To do this, an optimization algorithm (simulated annealing), aimed at maximizing the consistency indicator, will be developed subsequently. The formula for calculating the corrected generic consistency indicator proposed in Limayem [8] is as follows:

$$I.C._{\text{Generic corrected}}^{a,b} = \frac{1}{m^2}\left((m^2 - m)I.C._{\text{Regression corrected}}^{a} + \sum_{i=1}^{m} I.C._{\text{Diagonal corrected},i}^{b}\right) \tag{11}$$

Where: $a \in \{0,1,2,3,4\}$ and $b \in \{0,1\}$.

Indeed, the values of a and b are related to the types of errors to consider, namely:

- For a = 0: no error is tolerated.
- For a = 1: inconsistencies, not accessible to the model, related to all opinions (given by the different evaluators) associated simultaneously with the same two elements i and j (r_{ijk} and its symmetrical value r_{jik}, $i,j = 1,2,\dots,m$; $i \neq j$; $k = 1,2,\dots,K$) are tolerated.
- For a = 2: misalignment errors concerning the symmetrical opinions of the same evaluator are tolerated.

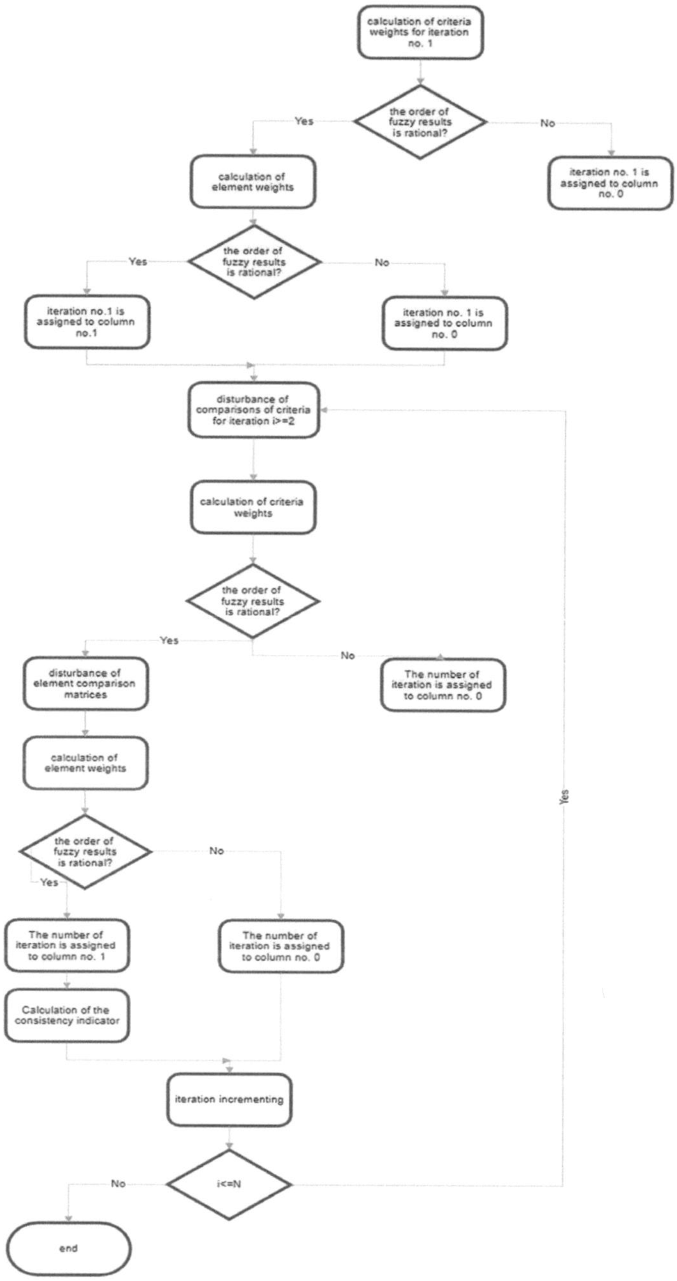

Figure 3.
Process of calculating weights and consistency indicators.

- For a = 3: misalignment errors concerning the symmetrical opinions of the same evaluator are not tolerated.

- For a = 4: pure errors concerning multiple opinions of the same comparison are tolerated.

- In our design project, we choose to opt for the following correction case: $I.C._{\text{Generic corrected}}^{0,1}$ because we consider that any inconsistency penalizes our consistency indicator, particularly the following two types of errors: misalignment errors and pure errors.

Therefore:

$$I.C._{\text{Regression corrected}}^{a} = I.C._{\text{Regression corrected}}^{0} = I.C._{\text{Regression corrected}} I.C._{\text{Regression corrected}}$$
$$= \frac{\sum_{i=1}^{m-1} \ln(\alpha_i)\left(\sum_{j\neq i}^{m}\sum_{k=1}^{d}\left(\alpha_{ijk}\ \ln(r_{ijk}) - \alpha_{ijk}\ \ln(r_{jik})\right)\right)}{\sum_{i=1}^{m-1}\sum_{j>i}^{m}\sum_{k=1}^{K}\left(\left(\alpha_{ijk}\ln(r_{ijk})\right)^2 + \left(\alpha_{ijk}\ln(r_{jik})\right)^2\right) - v_{\in}} i = 1, 2, \ldots, m,$$

Additionally, we have: $v_{\in} = v_{\in 0} = 0$ (in the case of a = 0). Moreover, using reciprocal comparison matrices, we have: b = 1 and $I.C._{\text{diagonal corrected}}^{1} = 1$ (according to the calculation formula proposed in the same study).

Finally, the adaptation of calculation Eq. (10) to the use of fuzzy input data gives:

$$I.C._{\text{Regression corrected}} = \left(I.C._{l}, I.C._{m}, I.C._{u}\right) \tag{12}$$

$$I.C._{l} = \frac{\sum_{i=1}^{m-1} \ln(\alpha_{il})\left(\sum_{j\neq i}^{m}\sum_{k=1}^{K}\left(\alpha_{ijk}\ \ln(r_{ijkl}) - \alpha_{ijk}\ \ln(r_{jiku})\right)\right)}{\sum_{i=1}^{m-1}\sum_{j>i}^{m}\sum_{k=1}^{K}\left(\left(\alpha_{ijk}\ln(r_{ijkl})\right)^2 + \left(\alpha_{ijk}\ln(r_{jikl})\right)^2\right)} \tag{13}$$

$$I.C._{m} = \frac{\sum_{i=1}^{m-1} \ln(\alpha_{im})\left(\sum_{j\neq i}^{m}\sum_{k=1}^{K}\left(\alpha_{ijk}\ \ln(r_{ijkm}) - \alpha_{ijk}\ \ln(r_{jikm})\right)\right)}{\sum_{i=1}^{m-1}\sum_{j>i}^{m}\sum_{k=1}^{K}\left(\left(\alpha_{ijk}\ln(r_{ijkm})\right)^2 + \left(\alpha_{ijk}\ln(r_{jikm})\right)^2\right)} \tag{14}$$

$$I.C._{u} = \frac{\sum_{i=1}^{m-1} \ln(\alpha_{iu})\left(\sum_{j\neq i}^{m}\sum_{k=1}^{K}\left(\alpha_{ijk}\ \ln(r_{ijku}) - \alpha_{ijk}\ \ln(r_{jikl})\right)\right)}{\sum_{i=1}^{m-1}\sum_{j>i}^{m}\sum_{k=1}^{K}\left(\left(\alpha_{ijk}\ln(r_{ijku})\right)^2 + \left(\alpha_{ijk}\ln(r_{jiku})\right)^2\right)} \tag{15}$$

The formula for calculating the consistency indicator given by formula No. (11) will thus be replaced by Eqs. (17)-(19):

$$I.C._{\text{Generic corrected},l}^{0,1} = \frac{1}{m^2}\left(\left(m^2 - m\right)\frac{\sum_{i=1}^{m-1} \ln(\alpha_{il})\left(\sum_{j\neq i}^{m}\sum_{k=1}^{K}\left(\alpha_{ijk}\ \ln(r_{ijkl}) - \alpha_{ijk}\ \ln(r_{jikl})\right)\right)}{\sum_{i=1}^{m-1}\sum_{j>i}^{m}\sum_{k=1}^{K}\left(\left(\alpha_{ijk}\ln(r_{ijkl})\right)^2 + \left(\alpha_{ijk}\ln(r_{jikl})\right)^2\right)} + m\right) \tag{16}$$

$$I.C._{\text{Generic corrected},m}^{0,1} = \frac{1}{m^2}\left(\left(m^2 - m\right)\frac{\sum_{i=1}^{m-1} \ln(\alpha_{im})\left(\sum_{j\neq i}^{m}\sum_{k=1}^{K}\left(\alpha_{ijk}\ \ln(r_{ijkm}) - \alpha_{ijk}\ \ln(r_{jikm})\right)\right)}{\sum_{i=1}^{m-1}\sum_{j>i}^{m}\sum_{k=1}^{K}\left(\left(\alpha_{ijk}\ln(r_{ijkm})\right)^2 + \left(\alpha_{ijk}\ln(r_{jikm})\right)^2\right)} + m\right) \tag{17}$$

$$I.C._{\text{Generic corrected},u}^{0,1} = \frac{1}{m^2}\left(\left(m^2 - m\right)\frac{\sum_{i=1}^{m-1} \ln(\alpha_{iu})\left(\sum_{j\neq i}^{m}\sum_{k=1}^{K}\left(\alpha_{ijk}\ \ln(r_{ijku}) - \alpha_{ijk}\ \ln(r_{jiku})\right)\right)}{\sum_{i=1}^{m-1}\sum_{j>i}^{m}\sum_{k=1}^{K}\left(\left(\alpha_{ijk}\ln(r_{ijku})\right)^2 + \left(\alpha_{ijk}\ln(r_{jiku})\right)^2\right)} + m\right) \tag{18}$$

From these last calculation formulas, we can clearly see that the consistency indicator I.C.i strongly depends on the initial comparisons: r_{ijk}, with i = 1, … ,m; j = 1, … , m; k = 1, … ,K.

3.2 Optimization algorithm

3.2.1 State of the art

Optimization problems have been studied for a long time, and several techniques have been used to solve them. Researchers from various fields have formulated their own optimization models depending on their specific needs. Generally, this involves analytically or numerically solving a model based on one or a set of variables to minimize or maximize. The quality of the results also depends on the choice of these variables, the relevance of the model, and the technical or computational support intended for solving the problem.

In the literature, there are different types of optimization problems, such as:

- Continuous optimization vs. discrete optimization
- Unconstrained optimization vs. constrained optimization
- Deterministic optimization vs. stochastic optimization
- Single-objective vs. multi-objective optimization

Moreover, Marco Antonio (Luersen [9]) discussed the functionality of optimization methods in treating optimization as part of global or local search. The optimization method comprises a global component (also called an explorer) that aims to explore the search space and a local component (called an exploiter) that exploits the resulting data (relationship between x and f(x)). In other words, the performance of the optimization method increases by integrating local search after global search and vice versa, through probabilistic resetting (depending on the considered objective) of the local search by considering the starting points and convergences achieved in each operation.

Thus, the optimization method we will use in this work, which aims to improve group coherence by maximizing the coherence indicator, will focus on searching for the global maximum without falling into a local maximum, thereby preventing the exploration of other solutions and limiting our optimization process. This method is known as *simulated annealing*.

3.2.2 Calculation algorithm

The simulated annealing method is an optimization algorithm inspired primarily by thermodynamics. It reflects the process of slowly cooling a previously heated metal, maintained at a high temperature to modify its characteristics. This process is repeated as many times as necessary for the set objective. The main characteristic of this method is that it is developed to avoid local minima and to search for the global minimum. The basic algorithm of the simulated annealing method is based on the following steps:

1. *Step 1: Definition of the problem:* This involves defining the initial work field.

2. *Step 2: Selection of algorithm parameters:* This step consists of setting the number of input data (starting points) for which it is necessary to calculate the energy (objective function), the initial temperature, the minimum or maximum temperature, the number of iterations, and the degree of temperature decrease while ensuring it is slow.

3. *Step 3: Measurement of system energy:* As the name suggests, this phase is dedicated to calculating the total energy from the initialized parameters.

4. *Step 4: Creation of fluctuations in the system:* This function allows the creation of disturbances relative to the starting points.

5. *Step 5: Implementation of the algorithm:* This step involves calculating the energies of the disturbances and comparing them with the previous results to find the optimal solution.

6. *Step 6: Cooling law:* In this calculation step, it is imperative to execute the temperature variation equation.

7. *Step 7:* Regeneration of the calculation loop.

4. Choice of architecture for the decision support platform: Overview of the specifications need validation

4.1 Validation of the need

The decision-making needs identified in both literature and the day-to-day management of projects have led to the development of this work. Indeed, the analysis of the state of the art in the first part resulted in the idea of designing a computing platform that overcomes the limitations and addresses various shortcomings. It is a platform intended for general use, allowing its parameters to be adapted to the characteristics of a given case.

Initially, the design project was approached by drafting specifications that outlined the main functionalities to be developed:

- *Functionality No. 1:* Calculation of weights of the data to be classified;

- *Functionality No. 2:* Creation of a classification table (with two columns, 0 and 1) corresponding to 1000 calculation iterations to classify the results concerning the order of fuzzy elements (column 1 if the order is respected, column 0 otherwise) and calculation of the equivalent coherence indicator;

- *Functionality No. 3:* Maximization of the coherence indicator (simulated annealing).

The consolidation of these three functionalities represents an innovative solution compared to the literature, in terms of both proposed methods and avoided dysfunctions. This technical solution will be implemented using MATLAB software. Graphical interfaces that materialize these functionalities will also be designed to facilitate user tasks and allow the visualization of intermediate results by offering the possibility to modify certain parameters (control performance).

4.2 Specification

The specification section aims to exhaustively describe the content of each key phase and the conditions for its validation. The execution of the following technical solutions (in the order used) will ensure the three aforementioned functionalities. These solutions include:

1. *Design of an interface for inputting linguistic values:* This interface allows users to input pairwise comparisons of data to be classified in linguistic terms (see **Table 1**).

2. *Development of a program for transforming linguistic values into fuzzy numbers:* This program converts linguistic values entered by users into their equivalent fuzzy numbers.

3. *Programming the algorithm for calculating criteria weights from a system of normal equations:* This algorithm calculates weights based on a system of normal equations (see formula (9)).

4. *Programming the calculation of the global weight of each data point to be classified:* This involves calculating the global weight of each data point (see formula (9)).

5. *Programming the calculation of the coherence indicator:* This program calculates the coherence indicator (see formulas (17)–(19)).

6. *Programming a set number of calculation iterations:* 1000 iterations are programmed for the test phase.

7. *Programming a two-column table to classify iterations:* This table checks whether the order of fuzzy element weights is respected (column 1 if the order is respected and column 0 otherwise).

8. *Programming an optimization algorithm:* The objective function of this algorithm is to maximize the coherence indicator.

9. Displaying the results corresponding to the maximum coherence indicator.

4.3 Design

4.3.1 Definition of the structure

Before implementing the calculation programs, we developed a reflection on the overall architecture of the platform. Moreover, displaying several interfaces is crucial for showcasing the development work behind it and improving the platform's attractiveness for broader use. The diagram below (see **Figure 4**) illustrates these interfaces within the decision-making process envisioned in this work.

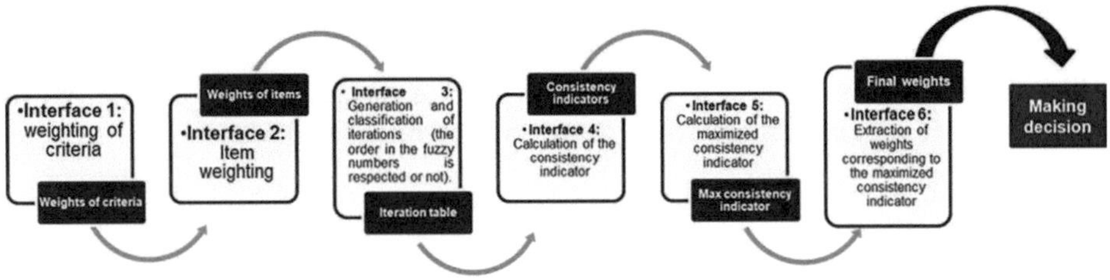

Figure 4.
Decision-making process in the platform.

4.3.2 Description of functions

The design phase aims to describe the "how" by detailing all technical and graphical functions (interfaces) that ensure the materialization of the functionalities. To this end, all functions developed *via* MATLAB are presented (in order) in the following table (**Table 2**), specifying the inputs and outputs for each.

4.3.3 Presentation of interfaces

A set of interfaces has been developed to facilitate communication with users. To implement these interfaces in MATLAB, we used the Graphical User Interface Development Environment (GUIDE) module. The main steps followed are:

1. Create the interface using the GUIDE editor;

2. Generate a .fig file for the Figure and an .m file for the code;

3. Complete the .m file containing the code.

4.3.3.1 Criteria weighting

This phase involves selecting criteria and comparing them in a cross-sorting framework to calculate a weight for each criterion. This phase is carried out using the interface structured as follows (**Figure 5**):

1. Parameter setting zone: This zone allows the user to conFigure the program by entering the number of criteria to be weighted and then activating the "set" button.

2. Execution zone: This zone includes an "execute" button that runs the weighting program.

3. Decision search zone: This area has a text field to specify the decision to be searched for and a "display" button to display it.

4. Decision insertion zone: This consists of a table for inserting user decisions, a "Save Decisions" button to store them, and an "Add Decision" button to generate a new decision.

MATLAB function	Inputs	Outputs
Remplissage_matrice_linguistique.m	Number of evaluators	Linguistic matrices
Convert.fis	Fuzzy numbers + linguistic values	Fuzzy numbers equivalent to linguistic values
Remplissage_liste_decideurs.m	Number of evaluators + number of criteria or elements	Lower matrices + Modal matrices + Upper matrices
Fonction_lower.m	Number of evaluators + number of criteria + Lower matrices + Upper matrices	αil\alpha_{il}αil
Fonction_modal.m	Number of evaluators + number of criteria + Modal matrices	αim\alpha_{im}αim
Fonction_upper.m	Number of evaluators + number of criteria + Lower matrices + Upper matrices	αiu\alpha_{iu}αiu
Ponderation.m	Number of criteria + αil\alpha_{il}αil + αim\alpha_{im}αim + αiu\alpha_{iu}αiu	Final weights of criteria (correction proposed by Boender et al. [2])
Ponderation_beta.m	Number of elements + Weights of elements relative to criteria (calculated by Fonction_lower.m, Fonction_modal.m, Fonction_upper.m)	Final weights of elements relative to criteria (after correction proposed by Boender et al. [2])
Somme_poids.m	Number of criteria + number of elements + Final weights of criteria + Final weights of elements relative to criteria	Final weights of elements
IC_lower.m	Number of evaluators + number of criteria + Lower matrices + Upper matrices + Final weights of elements relative to criteria	Lower coherence indicator I.C.01I.C.01I.C.01 Generic Corrected lll
IC_modale.m	Number of evaluators + number of criteria + Modal matrices + Final weights of elements relative to criteria	Modal coherence indicator I.C.01I.C.01I.C.01 Generic Corrected mmm
IC_upper.m	Number of evaluators + number of criteria + Lower matrices + Upper matrices + Final weights of elements relative to criteria	Upper coherence indicator I.C.01I.C.01I.C.01 Generic Corrected uuu
Perturbation_criteres.m	Number of evaluators + number of criteria + Lower matrices + Modal matrices + Upper matrices	Final weights of criteria (after perturbation)
Perturbation_éléments.m	Number of evaluators + number of criteria + number of elements + Lower matrices of elements + Modal matrices of elements + Upper matrices of elements + Final weights of criteria (after perturbation)	Lower matrices of elements + Modal matrices of elements + Upper matrices of elements (after perturbation) + Final weights of elements relative to criteria (after perturbation)

MATLAB function	Inputs	Outputs
Initialisation_recuit_simule.m	—	Maximized lower coherence indicator I.C.01I.C.01I.C.01 Generic Corrected lll + Maximized modal coherence indicator I.C.01I.C.01I.C.01 Generic Corrected mmm + Maximized upper coherence indicator I.C.01I. C.01I.C.01 Generic Corrected uuu
Perturbation_recuit_criteres.m	Number of iterations + Lower matrices + Modal matrices + Upper matrices	Final weights of criteria (after simulated annealing perturbation)
Perturbation_recuit_candidats.m	Number of iterations + Number of evaluators + number of criteria + number of elements + Lower matrices of elements + Modal matrices of elements + Upper matrices of elements + Final weights of criteria (after simulated annealing perturbation)	Final weights of elements relative to criteria (after simulated annealing perturbation)

Table 2.
Main MATLAB functions.

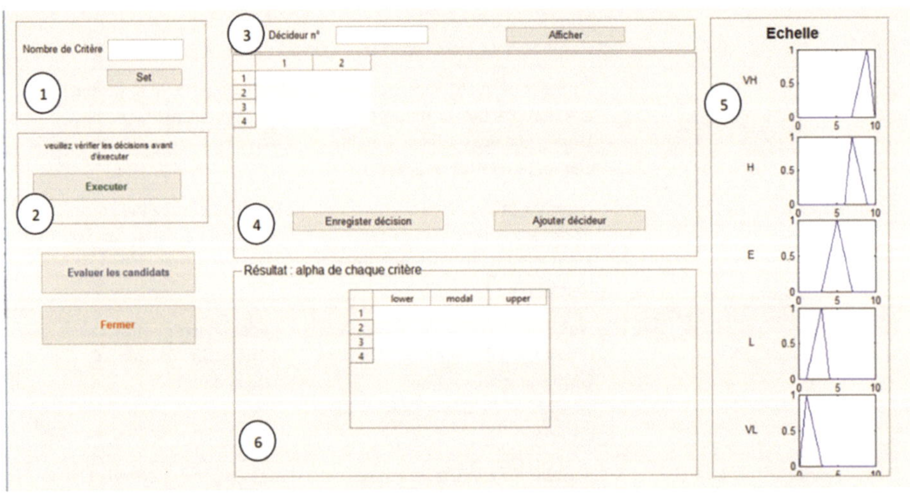

Figure 5.
Interface for criteria weighting.

5. Evaluation scale: Contains the linguistic values to be used for evaluation and their corresponding fuzzy numbers.

6. Result zone: This area displays the criteria weighting by showing three values: modal, upper, and lower.

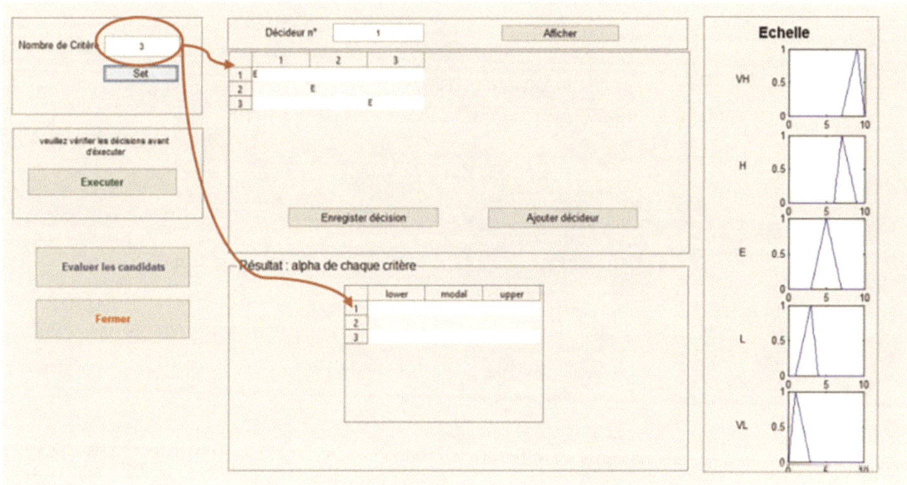

Figure 6.
Illustration of the criteria configuration step.

4.4 Step-by-step description of the criteria weighting process

- Step 1: Interface configuration

  This step involves setting the program by entering the number of criteria to be weighted. After this step, the table size in the decision insertion zone and the result zone adjusts according to the number of criteria (**Figure 6**).

- Step 2: Filling decision matrices

  This step follows the configuration and involves entering decisions and saving them. Only the upper part of the matrix is filled as it must be symmetrical; the comparison between two criteria is the same regardless of the direction. The diagonal is pre-filled with the "E" value, representing a neutral element since a criterion cannot be compared with itself. After entry, the decision is saved by pressing "save decision," and a message appears indicating the successful completion of the operation (**Figure 7**).

- Step 3: Adding new decisions

  To add another decision, press "add decision." A new matrix is generated, and the value in the display bar increments indicates the index of the current decision. This step is repeated for each decision-maker (**Figure 8**).

- Step 4: Verifying decisions

  In this step, simply enter the decision index in the search bar and press the "display" button to view the inputs for that decision (**Figure 9**).

- Step 5: Running the program

  After saving and verifying all decisions, execute the program by pressing the "execute" button. This yields the final weighting of each criterion as well as the upper and lower modal matrices for each decision (**Figure 10**).

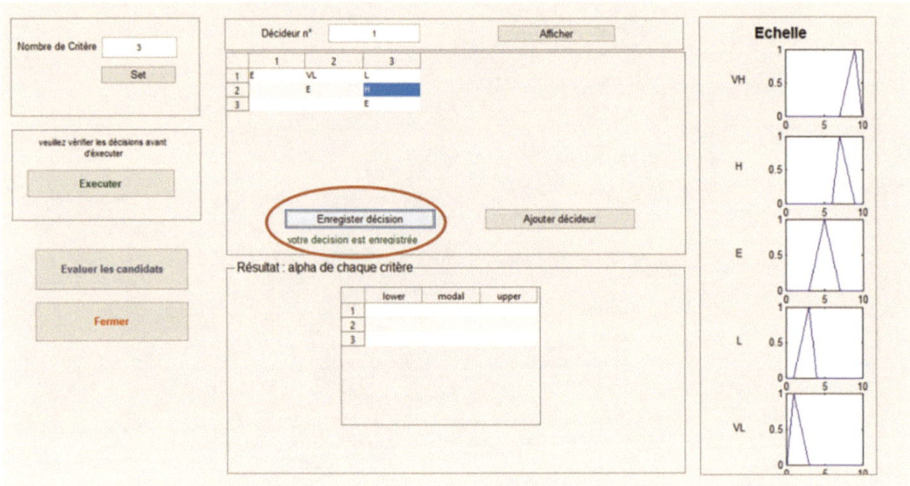

Figure 7.
Illustration of recording a decision.

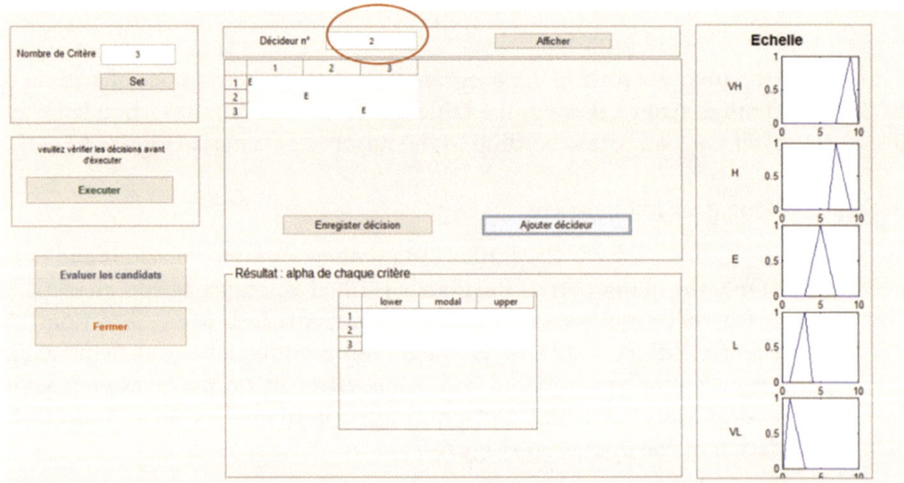

Figure 8.
Illustration of adding a new decision.

The first weighting phase concludes by pressing the "Evaluate Candidates" button to move on to the second phase or by pressing the "Close" button to end the session.

4.4.1 Weighting of candidates

This phase involves evaluating candidates based on each criterion and assigning a final weight to each candidate. This is done through the following interface (**Figure 11**):

The interface consists of several zones similar to those for calculating the criteria weights:

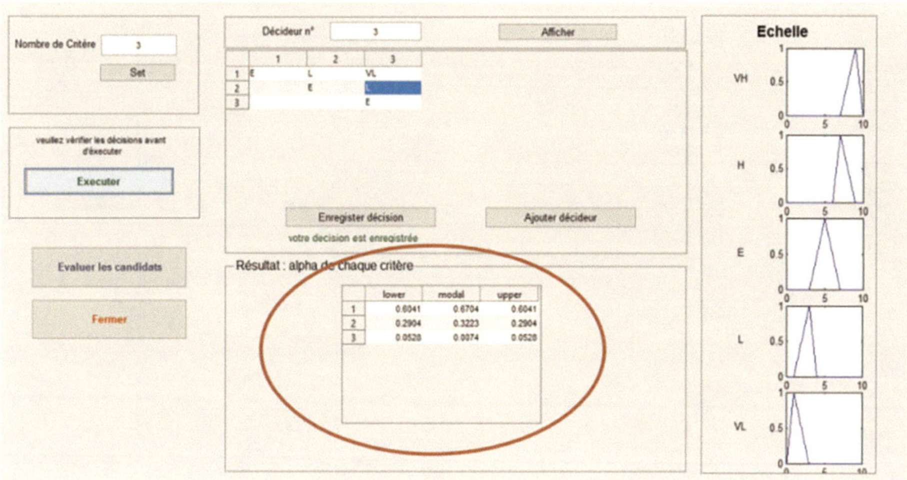

Figure 9.
Illustration of the execution and display step for criteria weighting.

Figure 10.
Display of modal, upper, and lower matrices.

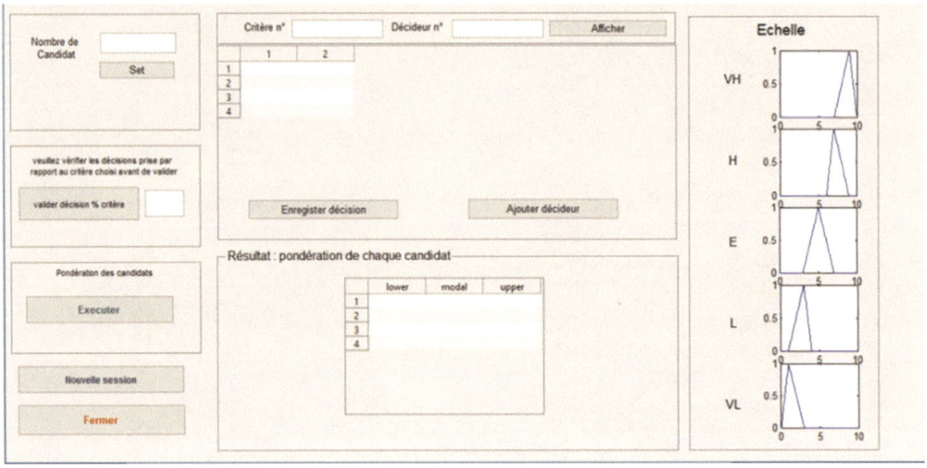

Figure 11.
Interface for candidate weighting.

1. Execution zone: Contains a button that allows the user to execute the weighting of candidates.

2. Evaluation scale: Contains the linguistic values used for evaluation and their corresponding fuzzy numbers.

3. Decision search zone: Allows the user to display a decision made regarding a specific criterion.

4. Decision insertion zone: Enables the entry of decisions about candidates.

5. Result Zone: Displays the weighting of the candidates.

6. Parameter setting zone: On this interface, the parameter setting is done according to the number of candidates to be evaluated.

4.4.2 Validation based on literature data

After setting up the architecture and completing the development part of our computing platform, we proceed to the crucial validation phase, which is essential for any design project. This is done using a series of calculation tests. For this, we use the example proposed in the work of Boender et al. [2], which describes the binary comparison matrix of three elements e1, e2, and e3 (see **Table 3**). In this example, we deduce the use of three rating values: E: (2/7, 1/3, 2/5), L: (1/6, 1/5, 1/4), and H: (2/5, 1/2, 2/3).

4.5 Problem of solving the weight calculation equation system

The initial tests of weight calculation using formulas (4) to (9) above, along with the resolution functions developed in MATLAB (Fonction_lower.m, Fonction_upper.m, and Ponderation.m), reveal a problem with calculating the inverse of the matrix (the matrix is not invertible). Therefore, to solve this problem, we opt to fix one of the weights by setting its three fuzzy values to arbitrary positive (non-zero) values. Then,

	e 1	e 2	e 3
e 1	(1, 1, 1)	(2/7, 1/3, 2/5)	(1/6, 1/5, 1/4)
e 2	(5/2, 3, 7/2)	(1, 1, 1)	(2/5, 1/2, 2/3)
e 3	(3/2, 2, 5/2)	(3/2, 2, 5/2)	(1, 1, 1)

Table 3.
Comparison matrix proposed in the work of Boender et al. [2].

	$(\alpha_{1l}, \alpha_{1m}, \alpha_{1u})$	(α2l, α2m, α2u)	(α3l, α3m, α3u)
Fixation of $(\alpha_{3l}, \alpha_{3m}, \alpha_{3u})$	(10.00, 10.95, 12.36) %	(26.64, 30.90, 36.59) %	(100.0, 100.0, 100.0) %
	(10.00, 10.95, 12.36) %	(26.64, 30.90, 36.59) %	(76.00, 76.00, 76.00) %
	(10.00, 10.95, 12.36) %	(26.64, 30.90, 36.59) %	(42.00, 76.00, 89.00) %
Fixation of $(\alpha_{2l}, \alpha_{2m}, \alpha_{2u})$	(10.00, 10.95, 12.36) %	(100.0, 100.0, 100.0) %	(49.8, 58.15, 66.5) %
	(10.00, 10.95, 12.36) %	(76.00, 76.00, 76.00) %	(49.8, 58.15, 66.5) %
	(10.00, 10.95, 12.36) %	(42.00, 76.00, 89.00) %	(49.8, 58.15, 66.5) %
Fixation of $(\alpha_{1l}, \alpha_{1m}, \alpha_{1u})$	(100.0, 100.0, 100.0) %	(26.64, 30.90, 36.59) %	(49.8, 58.15, 66.5) %
	(76.00, 76.00, 76.00) %	(26.64, 30.90, 36.59) %	(49.8, 58.15, 66.5) %
	(42.00, 76.00, 89.00) %	(26.64, 30.90, 36.59) %	(49.8, 58.15, 66.5) %

Table 4.
Weight calculation results for three fixing scenarios.

for a set of fixation scenarios, a sensitivity analysis of the results was performed to validate them. The table below presents the results of these calculation tests (see **Table 4**).

4.5.1 Comparison of results obtained by the platform with those in the literature

The use of the example from Boender et al. [2] aimed to compare the results of different cross-sorting methods using triangular fuzzy numbers, including:

- Bo.R.L.M.C (Logarithmic Regression Least Squares Approach): Proposed by Boender et al. [2].

- Bu.M.G.L (Original Buckley Approach Applied to the G.L.M Method): Refers to the Geometric Mean on Lines method (results provided in the work of Boender et al. [2]).

- F.W.A.M.G.L (Improved Fuzzy Weighted Average Algorithm by Scott for the Deterministic G.L.M Method): Proposed by Limayem [8] (results provided in the work of Limayem [8]).

- F.W.A.R.L.M.C.G (Improved Fuzzy Weighted Average Algorithm by Scott for the Deterministic R.L.M.C.G Method): Proposed by Limayem [8] (results provided in the work of Limayem [8]).

	Our developed platform	Bo.R.L.M.C	Bu.M.G.L	F.W.A.M.G.L	F.W.A.R.L.M.C.G
e1	(10.00, 10.95, 12.36)%	(10.5, 10.9, 11.8)%	(8.5, 10.9, 14.6)%	(8.7, 10.9, 14.1)%	(9.2, 10.9, 13.4)%
e2	(26.64, 30.90, 36.59)%	(28.0, 30.9, 34.9)%	(23.5, 30.9, 41.7)%	(25.4, 30.9, 37.8)%	(25.8, 30.9, 37.3)%
e3	(49.8, 58.15, 66.5)%	(49.8, 58.2, 66.5)%	(42.7, 58.2, 77.6)%	(50.4, 58.2, 64.4)%	(51.1, 58.2, 63.7)%

Table 5.
Comparison matrix proposed in the work of Limayem [8].

The results presented in **Table 5** detail the calculated weights of the three elements e1, e2, and e3 using our platform and those given by the respective authors. Indeed, the values obtained are very close, and the existing differences are related to decimal points.

5. Conclusion

Throughout this chapter, we have presented the different steps we followed to develop this decision-support platform. In fact, the need for design was identified and justified at the outset. Theoretical solutions were provided to address the limitations described in research work on decision-making performance. Next, we presented the theoretical approaches and algorithms adopted, including several calculations performed within the framework of cross-sorting methods and fuzzy logic. Classification and optimization algorithms are also proposed to improve consistency and accuracy.

Furthermore, the architecture of this decision-support tool was validated and described step by step. A detailed overview of the specifications, detailing the functionalities and interfaces intended for users, was provided. Finally, this computing platform must be validated to demonstrate its viability and theoretical contributions. For this purpose, a validation part using literature data was planned toward the end of the chapter.

Author details

Manal Tamir[1], Fatima Ouzayd[1]*, Chiheb Reddouane[1] and Hamid Ech-cheikh[2]

1 The Higher National School of Computer Science and Analyze Systems, Mohammed V University in Rabat, Morocco

2 Department of Naval Engineering, The Higher Institute of Maritime Studies in Casablanca, Morocco

*Address all correspondence to: fatima.ouzayd@ensias.um5.ac.ma

References

[1] Buckley JJ. Fuzzy hierarchical analysis, Fuzzy Sets and Systems, 1985; **17**(3):233-247. DOI: 10.1016/0165-0114(85)90090-9. ISSN 0165-0114

[2] Boender CGE, de Graan JG, Lootsma FA, Multi-criteria decision analysis with fuzzy pairwise comparisons, Fuzzy Sets and Systems. 1989;**29**(2):133-143. DOI: 10.1016/0165-0114(89)90187-5. ISSN 0165-0114

[3] Van Laarhoven PJM, Pedrycz W. Une extension floue de la théorie des priorités de Saaty. Fuzzy Sets and Systems. 1983; **11**:229-241

[4] Lootsma FA. Performance evaluation of nonlinear optimization methods via multicriteria decision analysis and via linear model analysis. In: Powell MJD, editor. Nonlinear Optimization 1981. Vol. 1, London: ed. Press A; 1982. pp. 419-453

[5] Escobar MT, Moreno-Jimenez JM. Distributions réciproques dans le processus de hiérarchie analytique. European Journal of Operational Research. 2000;**123**:154-174

[6] Vaisala. FDA, ICH: Regulations and Standards for Temperature-Controlled Supply Chains. Available from: www.vaisala.com/sites/default/files/documents/Cold_Chain_FDA_ICH-Application-Note.pdf

[7] Zadeh LA. Le concept de variable linguistique et son application au raisonnement approximatif. Information Sciences. 1975;**8**:199-249

[8] Frej Limayem. Modèles de pondération par les méthodes de tri croisé pour l'aide à la décision collaborative en projet. Sciences de L'ingénieur [Physics]. Français: Ecole Centrale Paris, 2001

[9] Luersen MA, Lopez RH. A Genetic Algorithm for Optimization of Hybrid Laminated Composite Plates. In: Muñoz-Rojas P, editor. Optimization of Structures and Components. Advanced Structured Materials. Vol 43. Cham: Springer; 2013. DOI: 10.1007/978-3-319-00717-5_4

Chapter 5

Electrical Machine Protection Relay Using Fuzzy Logic

Luis Benigno Corrales Barrios, Yainier Otero Rosell and Kirenia de la Caridad Viamontes Rabelo

Abstract

This article shows the development of comprehensive protection for electrical power machines, that is, motors, transformers, and generators, using fuzzy logic. The equivalent block diagram of any electrical machine to be protected is shown. The selection criteria for the protection of said equipment are shown, and the entire process necessary to apply fuzzy logic is explained. As an example, the development of the protection for a power motor is presented, and the results are compared with those obtained in other works by the authors and with other artificial intelligence techniques, which were adequately simulated, and it is observed that the protection operates under all possible failures and does not operate under nominal operating conditions.

Keywords: fuzzy logic, electrical machines, protection of power motors, protection of generators, protection of transformers, protective relaying, differential protection

1. Introduction

In artificial intelligence, fuzzy logic (LD) is used to solve a wide variety of problems, mainly those related to the control of complex industrial processes, decision systems, and the resolution of data compression.

The control implemented over fuzzy logic is one of the techniques used to implement protection techniques with controllers that imitate the human deductive process.

Advantages offered by this technique:

- It is conceptually easy to understand.
- It is flexible.
- Tolerates imprecise data.
- Can model nonlinear functions of any complexity.
- Make the most of the experience of experts.

- It can be combined with conventional control techniques.
- It is based on natural language.

One of the reasons for its wide use is that the conceptual basis of fuzzy logic rests on linguistic variables, that is, variables whose values are words instead of numbers.

Another reason, which plays a fundamental role, is the use in most applications of implication rules of the IF-THEN type, or simply fuzzy rules. The calculations of these rules serve as the basis for what could be called Fuzzy Dependency and Command Language (FDCL). In this connection, it is important to recognize that a solution in fuzzy logic is nothing more than the extrapolation of human solutions to FDCL because practically all human reasoning and concept formation are linked to the use of fuzzy rules. If we provide ourselves with a semantic structure for the computation of these rules, fuzzy logic greatly amplifies the possibilities of the mathematical treatment of human reasoning in Ref. [1].

2. Diagram for the protection of electrical machines

The machine to be protected (motor, generator, or transformer) is connected to the system as shown in **Figure 1**.

In the diagram in **Figure 1**, it can be seen that the machine is powered by a breaker. The machine's load current is minimized to values less than 5 A, through current transformers (CT). These pass the signal to a data acquisition card (DAC) and from there to the computer (COMP), which is responsible for processing it. When the protection must operate due to any damage to the machine, the computer returns a signal that the breaker must operate and the machine is taken out of service.

The protection of electrical machines is carried out as follows [2]:

a. Overload protection:

This protection must operate when the current exceeds the nominal value of the equipment to be protected (motor, generator, or transformer), according to the following expression:

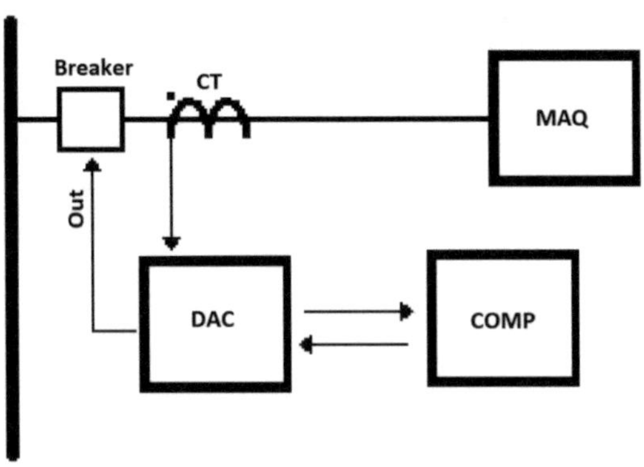

Figure 1.
Diagram of the machine to be protected [2].

$$I_r = K\, I_{máx\,load} \tag{1}$$

where: I_r is the relay setting current.

$I_{max\ load}$ is the maximum load current of the equipment to be protected.

K is a safety factor that takes into account the overload allowed in the equipment to be protected.

For the three machines, it is taken between 1.05 and 1.25 of the maximum load value.

The operating times of the relays are adapted to the overload characteristics of each equipment. For motors and generators, the inverse time characteristic is used, which is described in the following equation:

$$T = \frac{A\,D}{M^N - C} + B\,D + K \tag{2}$$

where:

D is the value of the time lever (generally from ½ to 10, although in some relays it reaches 12).

M is the value of the multiple (relationship between the real current and the relay setting current, M = I/Ir).

The rest of the coefficients are taken according to the values of the constants associated with each type of curve. In this case are:

A = 8.9341; B = 0.17966; C = 1.000; N = 2.0938; K = 0.028.

For transformers, the same value is taken as for motors and generators, but it only gives an alarm and the protection should not operate unless the overload is maintained for more than 2 hours.

b. Short circuit protection:

For small motors, generators and transformers, overcurrent protection is used:

$$I_r = K\, I_{ajus} \tag{3}$$

For transformers, the I_{ajus} is taken as the maximum short circuit current at its output and K = 1, for motors K is taken from 2 to 10 times the maximum load current and for generators, it is taken according to the short circuit levels at their terminals.

In all three cases, the operating time must be 0.1 second or less.

c. Differential protection:

This protection is based on taking the input and output currents of the windings of each machine, as shown in the following equation:

$$I_r = I_I - I_{II} \tag{4}$$

where I_I is the current through one part of the coil of each machine and I_{II} is the current through the other part of the winding, as shown in **Figure 2**.

As can be seen, the currents through the winding (represented by a black box) circulate normally according to the directions shown, so the current through the relay is zero (according to Kirchoff's first law). When there is a fault in the winding, one of the currents is reversed and a large current circulates through the relay.

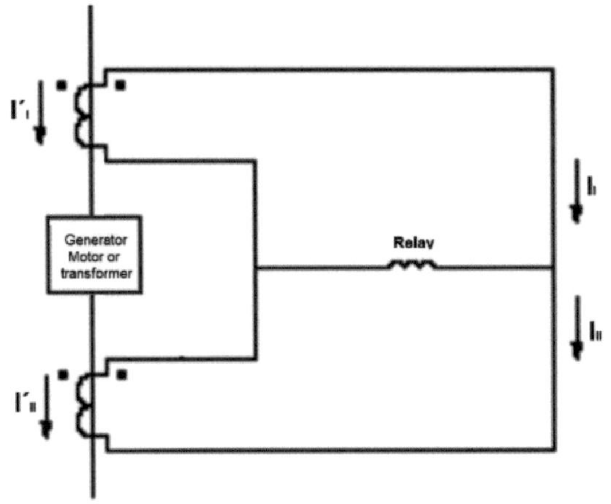

Figure 2.
Differential protection.

We clarify that this is the simplest type of differential protection and that there are circuit arrangements to ensure that the currents are in phase in the first place and, in most cases, differential percentage relays are used [2].

3. Argumentation of the rules used in this work

For each machine were selected the corresponding rules for the application of fuzzy logic, according to the protection criteria. These are described below:

a. Motors

High-power asynchronous motors are basically disconnected for two reasons: excessive heating of the motor windings and high current values produced by short circuits or rotor blocking; operations that are carried out using thermal protectors and any type of protection against short circuits [2, 3]. In this work, only failures or anomalies caused by the increase in current are taken into account.

The variables that directly influence the behavior of the protection against overload, over current (short circuit) and differential have been selected, so that the intelligent protector can make a decision: continue working normally or trip the breaker.

The following protection criteria were selected [4]:

Criterion C1

When the current take values equal to or greater than 1 (according to a previous setting), we are in the presence of an overload. For this fault, using the inverse time characteristic, the total time to wait for the relay to operate is calculated.

Criterion C2

When the current takes values equal to or greater than four times (although it can be adjusted to other values, depending on the value of the starting current), the overcurrent state is reached, in this case, the relay trips instantly.

Criterion C3

When the difference in currents (differential current) takes values greater than 1, we are in the presence of an internal fault.

b. Transformers

The increase in the use of electrical energy has required the use of high-power transformers in the different parts of an electrical system, plants, substations, and consumers. This makes this equipment more expensive, which implies that they require adequate protection.

Not only must high-power transformers be protected, but despite these machines being the simplest and most reliable within the electrical system, they must be adequately protected regardless of size. The type of protection to be used on a transformer generally depends on its nominal power.

Although transformers, and even more so those with nominal power above 500 kVA, have a reduced number of failures compared to other equipment in the electrical system, we cannot ignore or minimize their protections, since a failure in a transformer would lead to take this equipment out of service and therefore a considerable part of the system's consumers would have to be left without power.

Protection is achieved with the correct combination of system design and protection devices necessary to economically satisfy the application requirements. Protection must be designed to include:

1. The protection of the transformer against harmful conditions occurring in the system to which the transformer is connected.

2. Protection of the electrical system against the effects of transformer failure.

3. The detection and indication of conditions occurring within the transformer that may cause damage or failure.

It has been determined by studies carried out in different countries and by experience in the operation of transformers and autotransformers that they are, within the electrical system equipment, the ones that present the least failures in a certain period, for example in a year. However, to eliminate the defect, they are the equipment that is out of service the longest when they suffer a failure. A study carried out showed that, for transformers with primary voltage less than 15 kV, there were 0.0076 failures/year for each transformer, however, these failures reported 253 hours out of service for each failure produced in the transformers.

The fundamental faults that affect transformers and autotransformers are internal short circuits between their windings, between them and the ground or in the transformer terminals, and short circuits between the turns that make up their winding. Among them, short circuits between the terminals and internal short circuits in the windings are more common. Internal faults between the windings are less likely, three-phase faults are less likely in three-phase transformers and autotransformers, and short circuits between phases are practically impossible when there are groups of single-phase transformers.

Protection against internal faults in transformers must take them out of service instantly in an interval between 0.05 and 0.1 s, to minimize the damage that occurs in the transformer windings when these occur [2, 3].

Criterion C1

When the current take values equal to or greater than 1 (according to a previous setting), we are in the presence of an overload. For this fault, using the corresponding inverse time characteristic, the total time to wait for the relay to operate is calculated.

Criterion C2

When the current takes values equal to or greater than x times, the failure state is reached; in this case, the relay operates instantaneously.

Criterion C3

When the difference in currents (differential current) takes values greater than 1, we are in the presence of an internal fault.

c. Generators

The deterioration of insulation in the stator and rotor windings is one of the fundamental problems in generators. This occurs due to short circuits between phases that are associated with high short circuit currents that produce an electric arc at the faulted point, which deteriorates the insulation and can damage the iron sheets in the stator. A similar thing happens in phase-to-ground faults, because this current circulates through the stator, and can also damage the insulation of the generator winding.

We can summarize that the faults for which generators must be protected are the following:

- Short circuit to ground in the generator.
- Short circuit between phases in the winding and its terminals.
- Short circuit between turns of the winding of one phase of the stator when there is more than one winding per phase.

To protect generators against faults between phases in the stator winding, overcurrent relays or differential relays are used.

In generators greater than 1 MVA power and with any voltage, for voltages greater than 5 kV. and any power and for greater than 2.2 kV and greater than 501 kVA, the second variant is used and in smaller generators the overcurrent protection, which is sometimes connected in a differential type scheme [2, 3].

Criterion C1

When the current take values equal to or greater than 1 (according to a previous setting), we are in the presence of an overload, although in the generator this is compensated with internal adjustments of the same, so the protection's mission is only to signal. For this fault, using the corresponding inverse time characteristic, the total time to wait for the relay to signal is calculated.

Criterion C2

When the current takes values equal to greater than x times, the failure state is reached; in this case, the relay operates instantaneously.

Criterion C3

When the difference in currents (differential current) takes values greater than 1, we are in the presence of an internal fault.

4. Development of this work

For reasons of space, we will only expose what is related to motors.

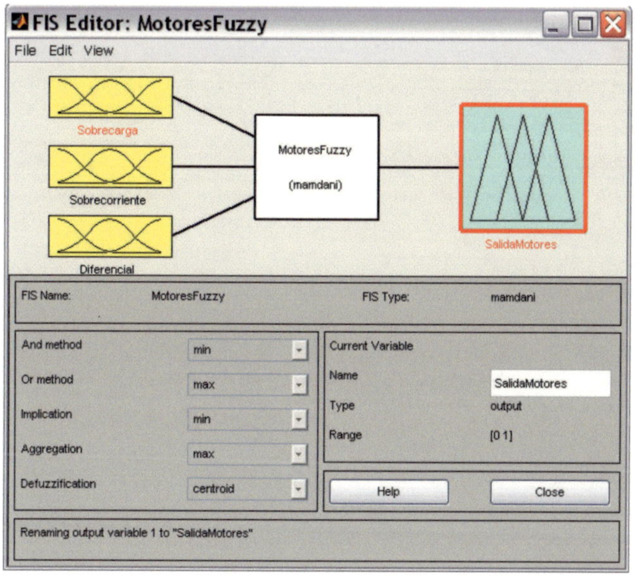

Figure 3.
FIS MATLAB fuzzy ToolBox editor.

The FIS Editor is the interface to use when creating a system based on fuzzy logic and presents general information related to the inference system, the names of each of the input and output variables as shown in **Figure 3** [5].

The Membership Function Editor is the interface that allows editing the membership functions of the inference system of the input and output variables as shown in **Figure 4**.

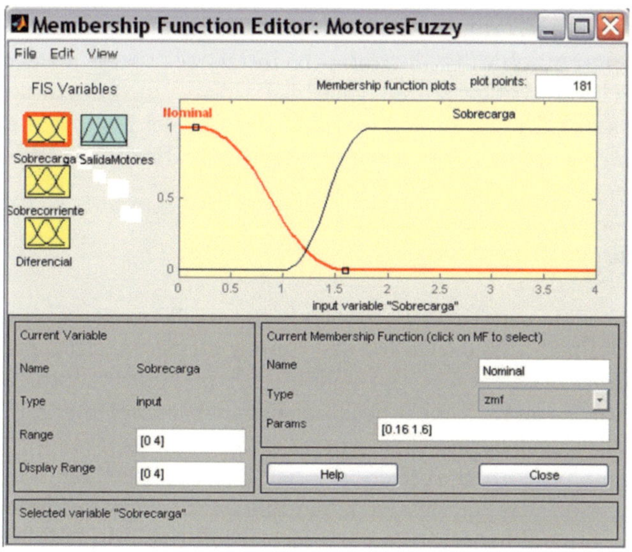

Figure 4.
ToolBox membership role editor.

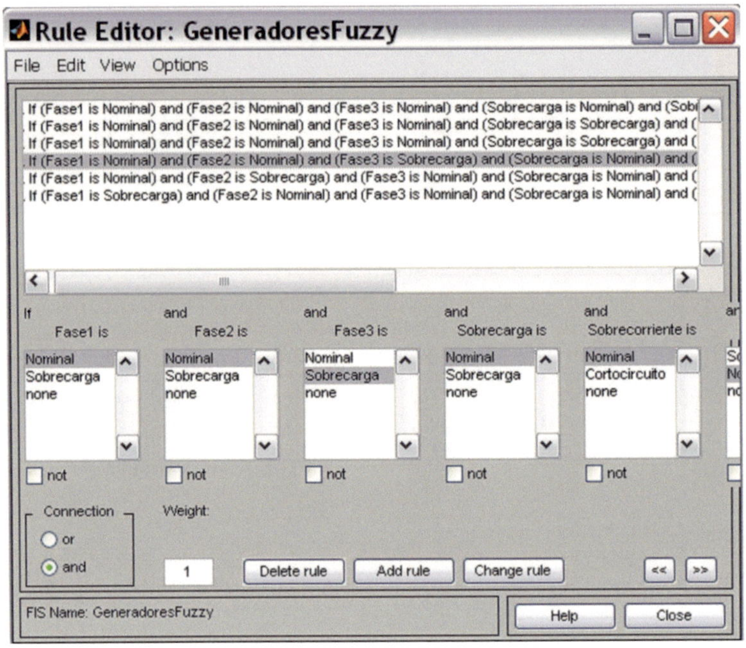

Figure 5.
ToolBox rules editor.

The rules editor contains a text field to present and edit the rules, in addition to the possibility of modifying the weight of each of them. The maximum value of the weights is one. This set of rules is assimilated by the computer as a matrix composed of a number of columns n, which corresponds to the number of input variables, another column corresponding to the output variable, another column that indicates the weight applied to each rule, and a final column where it is specified whether it is an OR rule (2) or an AND rule (1), as indicated in **Figure 5**.

The rules viewer graphically illustrates the inference process. Each rule is a row of graphs and each column of graphs is a variable.

In the fuzzy set of the output, the vertical line indicates the discrete value of this variable after having applied the corresponding defuzzification method as shown in **Figure 6**.

4.1 Fuzzification

Fuzzy Logic begins with the concept of a fuzzy set. A fuzzy set is one that does not present clearly defined boundaries and can contain elements with a partial degree of membership, unlike classical sets which completely include or completely exclude a given element, for example, X must be in set A or in its complement, the set NOT-A, that is, either in one or the other.

To illustrate more clearly the differences between both sets, let us take the example of thermal protection that is activated depending on the degree of heating of the windings of a motor and focus the analysis on the concept "admissible temperature during locked rotor failure in motors. 200 HP," whose values (determined by the operator) are in the range between 105 and 190°C.

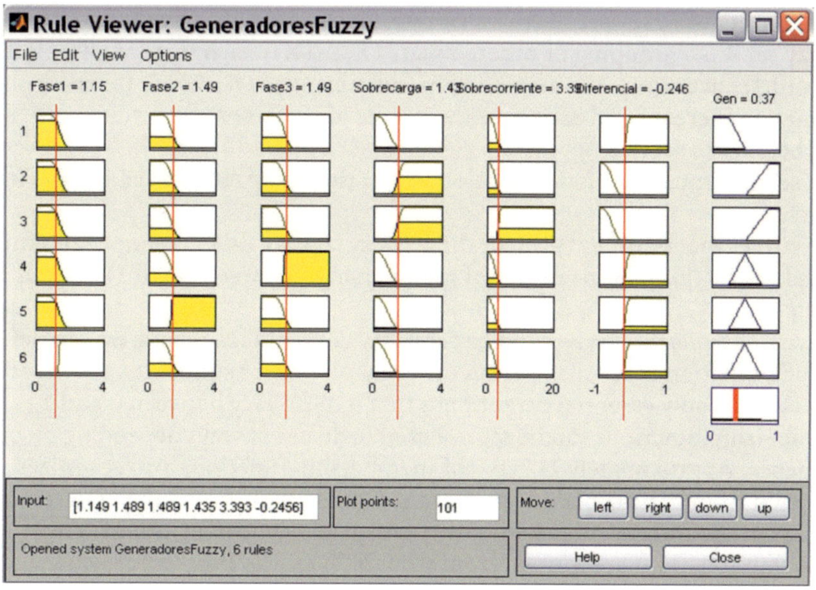

Figure 6.
ToolBox rules viewer.

Reasoning according to classical logic, (Boolean Logic) this concept can take only two values: true (with numerical value 1) for the temperature values $105 \leq T \leq 190$ and false (with numerical value 0) for the remaining values. According to fuzzy logic, it can take values such as 0.3 or 0.75, that is, it can take values in the interval between 0 and 1, including both terms. See **Figure 7a** and **b**.

As can be seen in **Figure 7a**, a temperature of 200°C according to Boolean logic (either one belongs to a set or one does not belong) has a value of 0, since it does not belong to the set "admissible temperatures of the stator," however, according to fuzzy logic, that temperature corresponds to an approximate value of 0.5 within the reference set, that is, its membership in that set is not completely ruled out. In terms of human language, it could be said that this temperature value is "a little permissible."

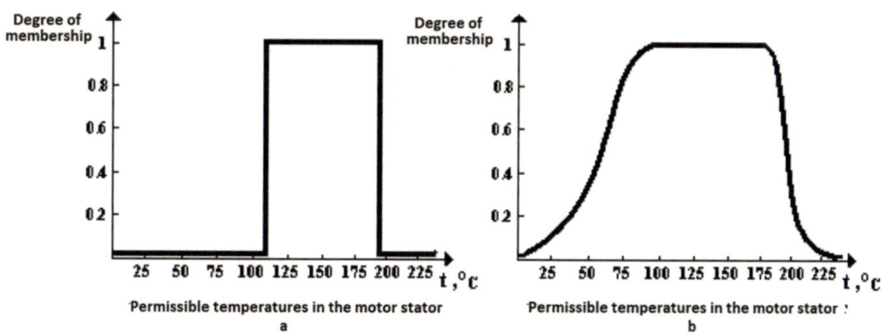

Figure 7.
(a) Values obtained according to classical logic. (b) Values obtained according to fuzzy logic.

The concept of fuzzy set, introduced by Zadeh in 1965 [6], is defined as follows: the fuzzy set A is a grouping of ordered pairs (X, $\mu_A(X)$) such that for all $x \in R$ it holds: μ_A: R → [0,1]], were μ_A is the membership function of the fuzzy set (or the membership function thereof) and expresses the degree of certainty of the statement "element x belongs to set A."

The set of all possible elements belonging to the fuzzy set is called the Universe of Discourse.

The curves that represent each of these fuzzy sets are called membership functions and define the value or degree of membership (between 0 and 1) of each of its elements.

These functions themselves can be arbitrary curves whose shape we can define by means of a function that suits us from the point of view of simplicity, speed, efficiency, etc. The only essential condition is that it must vary between 0 and 1.

Membership functions can be represented in different ways depending on our convenience. Approximately 11 types of membership functions are recognized, constructed from basic functions such as the Gauss function, sigmoid curves, and the curves of the square and cubic polynomials respectively, among others. The simplest membership functions are formed from straight lines, the simplest of which are triangular and trapezoidal. See **Figure 8a** and **b**.

Among the most used membership functions, we find bell-type functions, those described by Gauss distribution curves and curves that characterize situations of asymmetry such as sigmoid curves. See **Figures 9** and **10** respectively.

The fuzzification process consists of taking the inputs from the discourse universe and determining the degree of membership to each of the selected fuzzy sets using the membership functions. The input is always a numerical value limited to the universe of discussion of the input variable and the output, of this stage, is a degree of membership of the linguistic set (always the interval

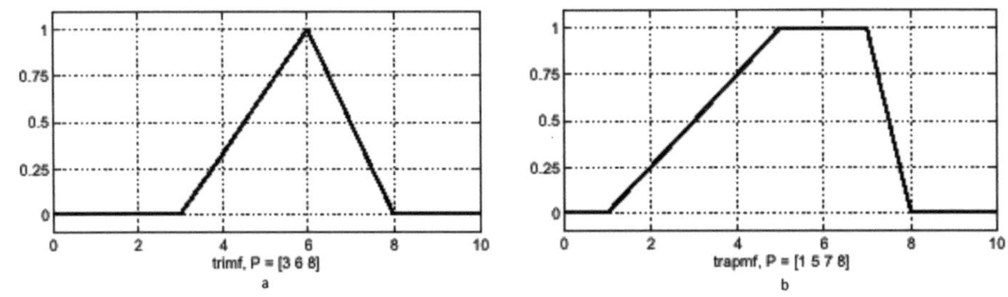

Figure 8.
(a) Triangular membership function. (b) Trapezoidal membership function.

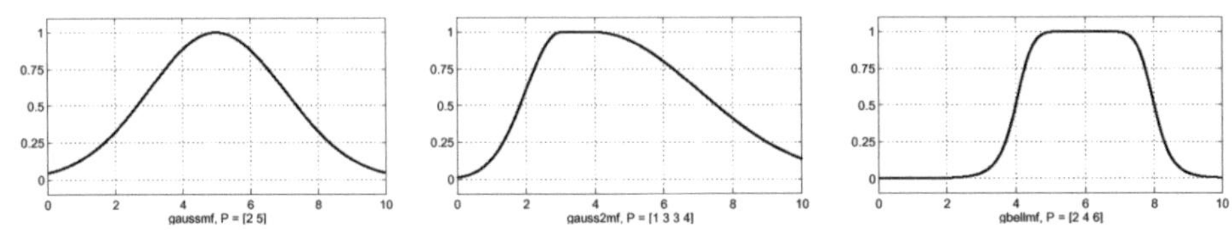

Figure 9.
Gaussian and bell-type membership functions.

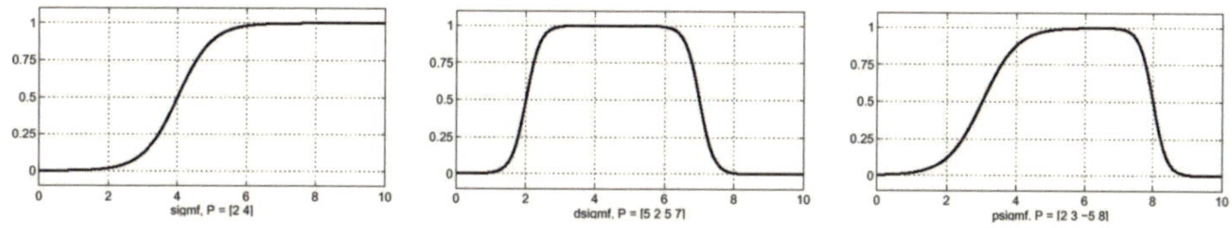

Figure 10.
Sigmoid membership functions.

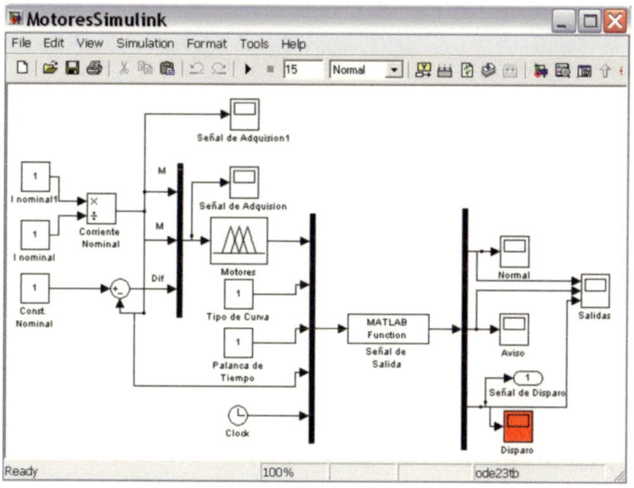

Figure 11.
Simulator for power motor protection.

between 0 and 1). The fuzzification of the input is performed based on the chosen membership function.

One of the most important aspects of logical reasoning within fuzzy logic is operations with fuzzy sets. Three basic logical operations are distinguished, similar to the AND, OR, and NOT operations of Boolean logic.

The "min" operation solves the relationship A AND B where A and B are functions limited in the range (0, 1) by using the function:

Min(A,B).

Similarly, using the same reasoning we can replace the OR operator with the "max" function so that the operation A OR B becomes:

max(A,B),

Finally, the NOT operation becomes its equivalent 1-A.

The simulation of the protection of a high-power motor was carried out according to the following diagram (**Figure 11**):

During the test to demonstrate the result of the simulation, it is compared against the simulation carried out in the [7] carried out with the MATLAB ToolBox for neural networks, in addition to a simulation carried out without applying artificial intelligence [7, 8], against the system implemented in this work. The results of this system were the following (**Tables 1–3**):

Likewise, simulations were carried out for transformers and generators.

As can be seen, the results of this work coincide with results obtained in others with different artificial intelligence techniques.

Overload Ir/In	Differential	Output signal normal	Output signal warning	Output signal trip
1.05	−0.05	1	0	0
1.1	−0.1	1	0	0
1.15	−0.15	1	0	0
1.2	−0.2	1	0	0
1.25	−0.25	1	0	0
1.3	−0.30	0	1	1
1.35	−0.35	0	1	1
1.4	−0.40	0	1	1
1.45	−0.45	0	1	1
1.5	−0.5	0	1	1
1.55	−0.55	0	1	1
1.6	−0.6	0	1	1
1.65	−0.65	0	1	1
1.7	−0.7	0	1	1
1.75	−0.75	0	1	1
1.8	−0.8	0	1	1
1.85	−0.85	0	1	1
1.90	−0.9	0	1	1
1.95	−0.95	0	1	1
5	−1	0	0	1

Table 1.
Test result for motors using fuzzy logic.

Overload Ir/In	Intelligent protection signal
1.05	0
1.1	0
1.15	0
1.2	0
1.25	0
1.3	0
1.35	0
1.4	0
1.45	0
1.5	1
1.55	1
1.6	1
1.65	1
1.7	1
1.75	1
1.8	1
1.85	1
1.9	1

Overload Ir/In	Intelligent protection signal
1.95	1
5	1

Table 2.
Test result using neural networks [8].

Overload Ir/In	Intelligent protection signal
1.05	0
1.1	0
1.15	0
1.2	0
1.25	0
1.3	0
1.35	1
1.4	1
1.45	1
1.5	1
1.55	1
1.6	1
1.65	1
1.7	1
1.75	1
1.8	1
1.85	1
1.9	1
1.95	1
5	1

Table 3.
Test result without using fuzzy logic [9].

5. Conclusions

1. An application was created with fuzzy logic that discriminates between nominal and fault conditions for electrical machines (motors, transformers, and generators) in cases of protection against overload and short circuits (overcurrent and differential).

2. It is verified with different typical cases that the developed application operates as protection of electrical machines, performing prediction functions and with the possibility of storing the different operating conditions.

3. The application developed is suitable using MATLAB's Simulink, which gives all the possibilities of making adjustments to possible relays and also constitutes future protection when a data acquisition card is added to the computer.

Author details

Luis Benigno Corrales Barrios[1*], Yainier Otero Rosell[2]
and Kirenia de la Caridad Viamontes Rabelo[1]

1 Universidad de Camagüey, Camagüey, Cuba

2 Dirección provincial de salud, Las Tunas, Cuba

*Address all correspondence to: luis.corrales@reduc.edu.cu

References

[1] Mukaido M. Fuzzy Logic for Beginners. Meiji: Work Scientific Meiji University, Japan First Publishing; 2001, reprinter 2002 and 2004. ISBN: 901-02-4534-3

[2] Corrales Barrios L. Protection of Electrical Machines. Camagüey: Editorial Universitaria; 2018. 159 p. ISBN 978-959-16-3012-4

[3] Altuve Héctor J, Schwitzer Edmund O. Soluciones modernas para protección, control y supervisión de sistemas eléctricos de potencia. Washington, USA: SEL; 2010. 430 p. ISBN-10: 0972502645, ISBN-13: 978-0972502641

[4] Morales JC. Inteligencia artificial aplicada a la protección de motores asincrónicos trifásicos. [Documento electrónico de la asignatura Inteligencia artificial de la carrera de Ingeniería eléctrica de la Universidad de Camagüey, Cuba]. Camagüey: Universidad de Camagüey; 2017. 60 p

[5] Jang R, Gulley N. Fuzzy Logic Toolbox. Ver.1. Massachusetts: The Math Works, Inc.; 1997. 208 p

[6] Goguen J. Fuzzy sets. Information and control, vol. 8 (1965), pp. 338-353.-LA Zadeh. Similarity relations and fuzzy orderings. Information sciences, vol. 3 (1971), pp. 177-200. The Journal of Symbolic Logic. 1971;**38**(4):656-657

[7] Alonso Alonso S. Application of neural networks for the prediction of failures in electrical machines [Degree thesis, Electrical Engineering]. Camagüey, Cuba: University of Camagüey; 2009

[8] Corrales LJCM, Escobar MA, Álvarez Y. Application of fuzzy logic to the protection of high-power motors. In: CD-ROM SAAEI-2001 Event Memories. Valencia, Spain: Universidad de Valencia, España; 2001

[9] Morales JC. Protection of electric motors [Masters thesis, Electrical Engineering]. Camagüey, Cuba: University of Camagüey; 2003